Prehistory

Guy Brett:
8.93

Prehistory

Giovanni Pinna

Illustrated with photographs
by Giovanni Pinna and Luciano Spezia

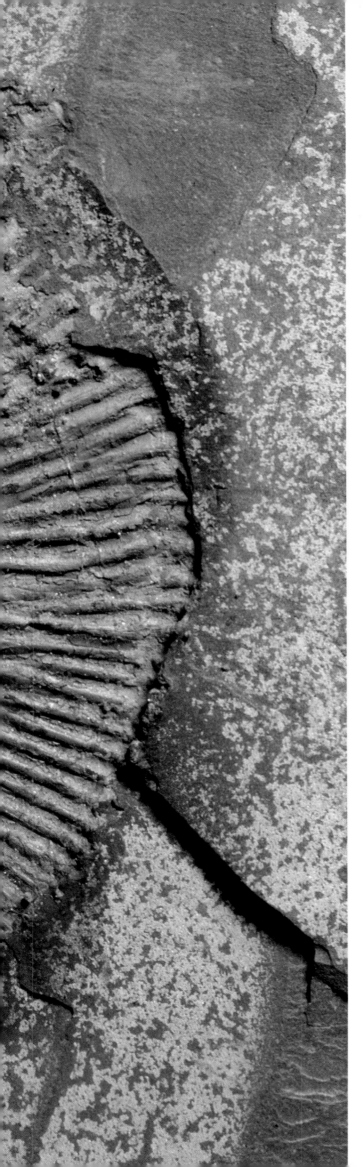

Contents

The Prehistoric World

  8 A Brief History of Palaeontology
20 The Death of Organisms
30 The Process of Fossilization
44 Rocks
54 Palaeoecology
72 Palaeogeography
78 Evolution

From the Beginning to the First Man

 97 Geological Eras
102 The Palaeozoic Era
128 The Mesozoic Era
164 The Cenozoic Era
178 The Quaternary Era

Extinct Animals

193 Classification of fossils

214 Glossary

217 Index

A Brief History of Palaeontology

Geocosmi Structura,
*a section of the Earth in a
seventeenth-century map*

"Shells can be found in the soil and on the mountains, and it is said that near Syracuse, in the Latomias, impressions of fishes and seals have been found, and on the island of Parus the impression of a bay leaf was found on a stone, and on Malta the tails of many species of marine animals." These words by Xenophanes of Colophon are the first known account of fossils being found in the Mediterranean basin.

Today we are no longer surprised by fossils; nobody wonders at the more or less petrified remains of ancient organisms which are found everywhere — seashells appearing within rocks at great distances from the sea; or the gigantic skeletons of animals long since vanished. The discovery of a mammoth's tooth in a gravel pit outside Milan is treated as the commonplace event it actually is.

Man has learnt about the past, has learnt the history of his planet. We have discovered the secrets which determine the constant evolution of the organic world, and we are no longer surprised when the traces of this evolution and of this past appear on our doorstep. However, this was not always the case; even though fossils — the remains of ancient history — have surfaced in all parts of the world since antiquity.

The history of the Earth, with the problems of its origin and development, have always interested man, just as fossilized organisms have always elicited the curiosity of naturalists and philosophers. Palaeontology is not an ancient science; it can be said to have begun only in the eighteenth century. However, even before this date, the traces of the Earth's ancient history were observed and records of discoveries have been handed down to us either in legends and poems, or in scientific and philosophical treatises.

Ever since the earliest times, man has felt the need to explain the presence of fossils. The priests of ancient Egypt acknowledged that various generations of animals had lived on the face of the Earth, following in one another's footsteps in a long chain of cataclysms and creations which strongly resembles the theories of Baron Cuvier, the founder of modern palaeontology. The priests' ideas are understandable in the light of certain fossils which have been unearthed at Fayum (in Egypt) during this century. Strange animals were found which had lived between sixty and thirty-five million years ago: elephants the size of pigs; others with four tusks; and mammals resembling rhinos (such as the *Arsinoitherium*) with two thick horns on their forehead.

Classical mythology is often a more abundant record of fossils than the rocks themselves. According to all creeds, man was born out of the earth. This fertile mother which produces wheat, maize and rice, the basis of most civilizations, also produced fossils. They are known today for what they are but in former times they were probably the origin of many a legend. Thus, in ancient mythology Cadmus' race was born out of the earth; the giants of Hesiod were the children of Rhea; and Enkidu (one of the characters in the Mesopotamian saga of Gilgamesh) was moulded out of clay like Adam.

During Greek and Roman times, fossil findings caused varying reactions: some people guessed their meaning, others resorted to abstruse theories. Xenophanes, for instance, understood that the sea must have once covered

10

Four engravings from the book Hortus Sanitatis *by Johann Wonneck von Cube (Venice 1511), representing fantastic and monstrous animals*

the dry lands where the remains of marine animals were found. According to Ovid, Pythagoras reached the same conclusions, and so did Herodotus, the historian who can be considered the first geologist. He wrote: "I believe Egypt to have been just such a gulf, stretching from the northern sea toward Ethiopia ... I believe those who maintain this ... having noticed that shells appear on the mountains." This actually amounts to the first palaeogeographical reconstruction to be based on fossils, and it is a fairly correct one: towards the end of the Mesozoic and the beginning of the Cenozoic, some sixty million years ago, the sea partly covered Egypt and left behind, on the rocks at the edge of the desert, the shells of gastropods and lamellibranchs or tiny nummulites, the round protozoa which form the rocks used in the construction of the Giza pyramids. An ancient historian, Strabo, tells how these nummulites were, in his days, regarded as the petrified left-overs of the food eaten by the workers who built the pyramids: in effect, fossilized lentils. Herodotus, that tireless traveller, when writing about Arabia, recounts: "I saw an indescribable quantity of bones of serpents and of dorsal spines; there were heaps of dorsal spines, some large, some less so, some small, and there were a great many of them." It was probably one of the fossil deposits which gave rise to the legend of winged serpents. Such findings abound in Africa: millions

The Giza pyramids were built with blocks of Eocene limestone containing large quantities of fossilized nummulites. The lenticular organisms were once believed to be petrified lentils, the remains of the food consumed by the workers who had built the pyramids

of years ago during the Mesozoic, the continent was ideally suited to the lifestyle of the large dinosaurs; lush plains and vast marshes offered them a perfect environment, rich in nourishment. The remains of these vanished reptiles often come to the surface in the desert sand. An ancient Tuareg legend, telling about stone serpents beyond the mountains, led a few years ago to the discovery of an incredible deposit in the Teneré desert, beyond the Aïr mountains: entire dinosaur skeletons were lying on the surface, their long backbones resembling elongated serpents. It is therefore quite possible that Herodotus may have found just such a fossil graveyard during one of his many journeys. If this was the case, he must this time have been induced by the legend to neglect the scientific aspect, something he did not do in the case of the shells discovered in Egypt.

Many other legends grew up from the discoveries of gigantic bones; some are famous, like the one of Polyphemus as told by Homer. The figure of Polyphemus, the giant with only one eye in the middle of his forehead, is certainly due to fossil findings. Elephants lived in Sicily during the Quaternary. They were smaller than the elephants we know today and smaller than those which lived elsewhere in Europe in the same period. The skull of an elephant has a particular structure and the most striking part of it is a hole in the middle of the forehead. The hole does not correspond with either of the eyes, it represents the opening of the nostrils and the root of the trunk. Such skulls are frequent in Sicilian Quaternary sediments as well as in those of many other Mediterranean islands, and it would have been quite natural for the ancient inhabitants to believe them to be the remains of gigantic one-eyed creatures. Homer also tells, in the Odyssey, how the sirens' beach was covered with the bones of unfortunate sailors who had been led to their deaths. There may well be a connection between this legend and the layers of bones which can be found in many of

An illustration from the book Mundus subterraneus *by Athanasius Kircher (Amsterdam 1678). The original caption reads: "Here is the four-footed winged dragon, well-known at all times, which Deodatus of Gozo, a Jerusalem Knight, captured on the island of Rhodes by the trick previously described; on account of his services to the island he was later elected Grand Master of the Order"*

the grottoes along the Mediterranean shores, the so-called ossiferous breccias formed by the accumulation of the remains of ancient animals. Sinbad the sailor, during one of his journeys, was shipwrecked and managed to save himself by reaching the beach of a lush island. Once he had recovered from the initial shock, the fearless sailor from the *Thousand and One Nights* went on to explore the island. He came to a large white dome. He walked all round this curious building and could find neither doors or windows; then suddenly the sky darkened and the great bird Roc appeared. Thus Sinbad realized that this large dome was nothing but the gigantic egg of that mythical creature. The tale, however incredible, is probably based on truth and shows once again how the discovery of fossils might in the past have contributed to the birth of fascinating legends.

Sinbad's journeys had probably taken him far south along the eastern coasts of Africa and he was shipwrecked on the shores of Madagascar. Fossils are plentiful in Madagascar, and among them are the remains of a flightless bird, a kind of ostrich, almost three metres high, called *Aepyornis*. Not only its bones have been found on the island, but also its eggs, often undamaged and seven times bigger than those of an ostrich. Arab merchants almost certainly knew Madagascar and it is quite possible that they may have found both the skeletons and the eggs of *Aepyornis*. Their tales about them would have grown more and more fantastic during the long days at sea.

Even in modern times, legends and tales are still being told which are undoubtedly connected with the appearance of fossils. In many alpine areas there are stories of the footprints of the devil being impressed on the rocks. The prints are indeed there and many have seen them but they have nothing to do with Lucifer. They are shells such as *Conchodon*, which lived 200 million years ago and can often be found in the rocks as a section in the exact shape of a horse's hoofprint.

A classic representation of extinct animals, probably an iguanodont and a megalosaur, in Rion's etchings for the book La terre avant le Deluge *by Louis Figuier (Paris 1863)*

Stories of witches, devils and dragons characterized the Middle Ages; the latter were so well-known that a proper zoological classification was invented for them. Many of them certainly had their origin in the marriage between popular imagination and the fossils which were suddenly unearthed, perhaps while ploughing a field. Such is the case of the dragon sculpted on a fountain in Klagenfurt in Austria. It probably represents the first attempt at reconstructing an animal from the past, in this case a woolly rhinoceros such as lived in Europe during the Quaternary. The skull of one such animal had been found near the town and is still housed in the local museum.

It is much the same with other dragons which were precisely named, like the *Draco bipes apteros captus in Agro Bononiensi:* a Bolognese dragon, which appeared in an area rich in fossils of vertebrates; or the stone serpents in the coat of arms of York, coiled serpents which are simply ammonites, plentiful in this part of Great Britain. A rudimentary head was sculpted onto one of these ammonites, now in the British Museum, to stress the image of a serpent.

Thus, throughout the history of palaeontology, accurate interpretations were often accompanied by fanciful ones. Empedocles of Agrigentum is

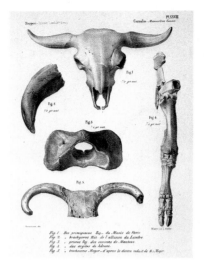

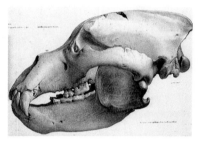

The skeleton of an Anoplotherium commune *found in the gypsum of Montmartre and illustrated in George Cuvier's book* Recherches sur les ossemens fossils *(1812), in which he laid the foundation for modern palaeontological research*

From top to bottom: *Three illustrations from Italian palaeontological books of the last century (written in French!): the first two are from Emilio Cornalia's* Mammifères fossiles de Lombardie *(Milan 1858–71) and show fossil remains of an aurochs and a cave bear. Below is a geological map of the Esino area from Antonio Stoppani's monograph* Les petrifications d'Esino *(Milan 1858–60)*

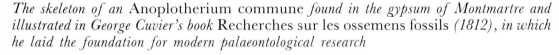
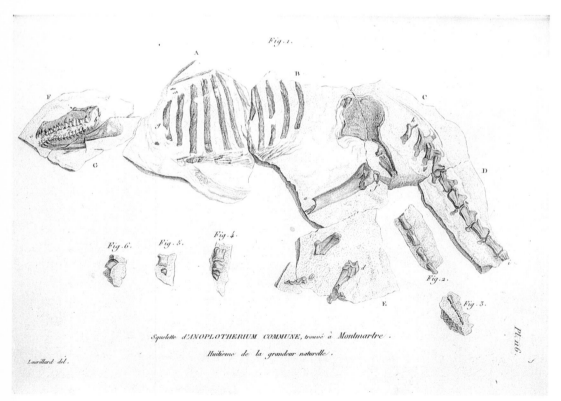
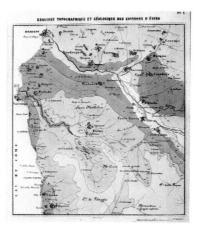

The Flood as seen by Francis Danby (1793–1861). For centuries the Flood was considered to be the reason why marine fossils were found on dry land

among those who got closer to the right diagnosis: he stated that fossils belonged to groups of animals very different from contemporary ones. Theophrastus of Ephesus acknowledged fossils to be the remains of organisms which had once been alive; and Pliny the Elder mentioned ammonites and nummulites, and he fully understood the liquid origin of amber — the solidified resin of Tertiary conifers — and consequently the very nature of the insects it contains.

It is surprising that Aristotle never mentions fossils in any of his numerous scientific and natural history works, even more so in the light of the theory which Avicenna attributes to him. On the basis of certain ideas which Aristotle outlined in his *De Respiratione*, according to which living organisms can be born out of the earth, Avicenna stated that an unidentified force was able to generate fossils within the rocks themselves.

The total lack of any Aristotelian pronouncements on the subject and Avicenna's abstruse theory obviously affected "palaeontological" thought for years to come. It is not until much later that Leonardo da Vinci, Fracastoro and Bernard Palissy bring us back to the idea of an organic origin of fossils. The fantastic theories which characterized palaeontological research come, at least partly, to an end during the last years of the sixteenth century, and with the seventeenth century an age of more scientific research was born.

In Italy, Fabio Colonna clearly distinguished marine fossils from terrestrial or freshwater ones and compared the teeth of fossil sharks with those of living ones. The Dane, Niels Stensen (Nicolaus Steno) first recognized the differences between marshy deposits and marine deposits and stated that the strata containing fossilized molluscs were formed under water and were shifted to their present position by subsequent geological events. The history of palaeontology begins with these ideas and from then on the number of scientists who regarded fossils as animals and plants from the past constantly increased.

Towards the end of the seventeenth century the organic origins of fossils were almost unanimously accepted and the theory that they were actually relicts of the Flood became prevalent. The theory itself has ancient origins and is based on the frequent findings of shells and other remains of marine organisms in areas which are at present very far from the sea. The Flood plays a major part in the biblical tradition, in Mesopotamian mythology, in Central American legends — all with very little variation on the theme and, until a short time ago, formed the central core of scientific interpretations. At the beginning of the eighteenth century, the Swiss Scheuchzer attributed to a child drowned in the Flood a fossilized skeleton found near Oehningen, Germany; later Cuvier recognized the skeleton as belonging to a large Miocene salamander, and it was called *Andrias scheuchzeri* in honour of its discoverer.

It was Cuvier who formulated the methods and aims of palaeontology thus making it into a modern science. His main work, *Ossemens fossiles*, laid the

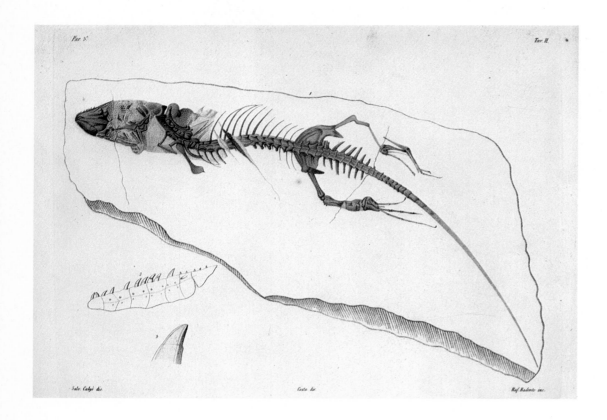

A rhynchocephalian reptile of the Cretaceous as illustrated in the book Paleontologia del Regno di Napoli *(1850) by the Italian palaeontologist Oronzo Gabriele Costa*

A caricature of the famous English palaeontologist Sir Richard Owen, riding a Megatherium skeleton. Owen, the first director of the scientific section of the British Museum, was a staunch antagonist of Darwin in the continuing argument on evolution

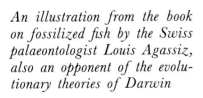

An illustration from the book on fossilized fish by the Swiss palaeontologist Louis Agassiz, also an opponent of the evolutionary theories of Darwin

17

A 1902 photograph of the famous Beresovka mammoth, found in Siberia encased in ice and complete with its skin and fur

foundations of modern comparative anatomy by enunciating the theory of correlation of organs. Cuvier also launched into serious reconstructions of extinct animals, often on the basis of a few incomplete remains.

Since Cuvier's day, the history of palaeontology is no longer based on fantastic legends, but on increasingly serious research and improving technology, and the general picture of our planet's past life has thus become gradually more detailed. The publication of Darwin's theory of evolution transformed palaeontology into one of the most important of natural sciences: not only was it at the basis of the theory of evolution, it also became the only science capable of providing tangible evidence of the past evolution of living creatures and of the wide-ranging transformation which took place within the animal and plant kingdoms during the millions of years of the Earth's history.

It has been scientifically established that fossils are either the remains of organisms which lived in the past or the traces of their existence. It follows that palaeontology is a science dealing with life itself, just like biology, botany and zoology. A palaeontologist has to regard fossils as if they were living matter, or at any rate has to take into account the fact that they once were.

This is actually one of the more recent innovations in the field. For years fossils had been regarded as inert matter, like minerals. This led researchers to stress the most minute differences and to attempt a detailed classification which did not take into account individual variations and all those characteristics which prevent any living organism from being identical to any other. The new attitude, which considered fossils as the remains of ancient populations of organisms with all their biological characteristics, represented a real scientific revolution. It was only at the beginning of this century that palaeontologists realized that they were dealing with live matter subjected to the same biological laws which affect the organisms of everyday life. This concept has now become the very foundation of palaeontology and should always be kept in mind if one is to achieve a realistic reconstruction of the history of life.

Fossilized dinosaur bones being retrieved in the Dinosaur National Park, Utah (United States of America). The public is admitted into the Park to visit the excavation work

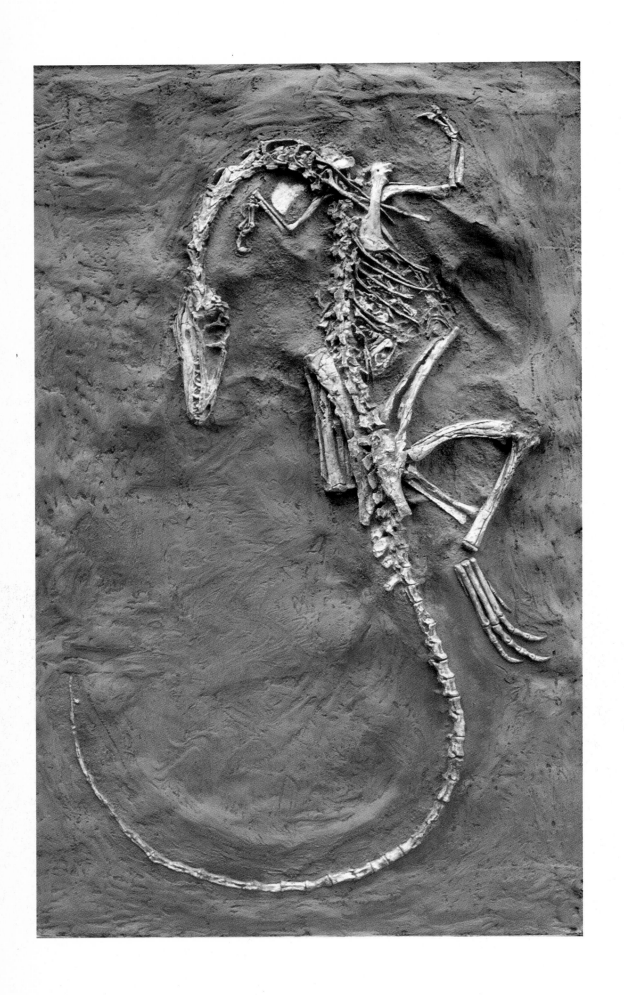

Coelophysis bauri: *the complete skeleton of one of the oldest known carnivorous dinosaurs, found in the deposits of the Upper Triassic in New Mexico. The backward curve of the neck is due to the stretching of the tendons which took place immediately after the death of the animal. Many fossilized skeletons are found in this posture*

The Death of Organisms

Dying is the last act in the life of an organism. After death, chemical, physical and biological processes take over which may lead to conservation, that is, to fossilization.
Palaeontological research begins therefore with the cause of death: knowing how and why an organism died allows the scientist to formulate hypotheses concerning the environment in which the creature lived and the relationship between environment and organism at the moment of death. The death of an organism, be it plant or animal, may be caused either by internal or by external factors; it can be either natural or violent.
Natural death obviously occurs infrequently in an animal world dominated by pitiless, competitive laws; an animal rarely dies of old age or illness. From a biological point of view, it is not particularly interesting to establish whether an animal or a plant died of old age; this would only indicate that a certain plant or animal had been lucky enough to survive the ruthless competition of other organisms around them, or else they were so tough as to be able to put up an effective resistance. It is also very difficult to determine whether a fossil's age was such as to indicate a natural, non-traumatic death. The age of constantly growing animals can be guessed by their size: one can, for instance, assume that the larger the shell, the older the animal was when it died. The age problem is rather more complicated in the case of organisms with uneven growth. In the case of arthropods for instance; since they grow through periodical moulting of their exoskeleton, age can only be estimated if one knows the yearly growth rate. Age is easier to determine in the case of vertebrates: certain fishes have been found to develop growth-rings on their scales for each year of growth, so that it is fairly simple to establish their age by counting such rings. Where mammals are concerned, an age estimate is even easier: senile specimens show a general decalcification of their skeleton, the disappearance of sutures between skull bones and, above all, the wearing out of teeth, particularly in the crown area.
Sometimes death has been caused by afflictions which have left a clear mark on the bones — the only part of the organism to survive in the

The embryo of an ichthyosaur (genus Stenopterygius*) perfectly fossilized within the deposits of the Lower Jurassic in Holzmaden, Germany. The sediments of the Holzmaden deposits, being very fine, have preserved very accurately the organisms of the past, among which were two ichthyosaur females which died in childbirth, their embryos and the newly born young*

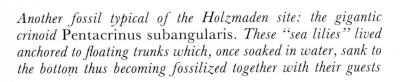

Another fossil typical of the Holzmaden site: the gigantic crinoid Pentacrinus subangularis. *These "sea lilies" lived anchored to floating trunks which, once soaked in water, sank to the bottom thus becoming fossilized together with their guests*

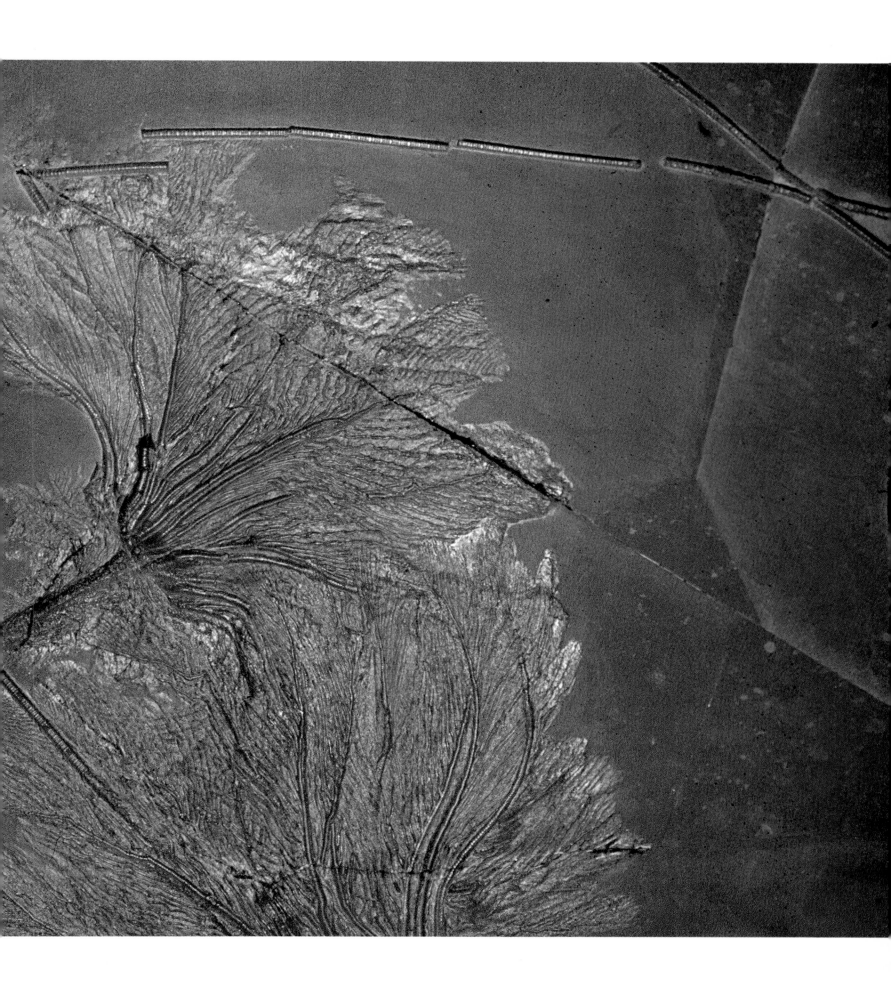

fossilized state. Occasionally, by analyzing these bones, it has been possible to establish the nature of the affliction. In some cases it has been noted that illnesses typical of man were also present in extinct vertebrates. Frequent among these were osteoarthritis, breakages and pathological phenomena such as abscesses, ulcers and sores affecting areas very close to the bones themselves.

Interesting examples of such afflictions have been discovered in fossil reptiles. Certain specimens of *Metriorhynchus*, a Jurassic crocodile, showed clear signs of tuberculosis and necrosis in their palate and pelvis bones. Specimens of Jurassic sauropods — gigantic quadrupeds — show evident signs of osteoarthritis and bone ankylosis. It is very likely that death was not actually caused by these afflictions, which are not very serious in themselves, but by the fact that the weakened organism would have partially lost its mobility and would thus have fallen prey to predators. Finally, the discovery of an unusual fossil of a marine reptile belonging to the Ichthyosauria has enabled scientists to formulate interesting hypotheses not only regarding its death but also, more generally, on the basic biology of these ancient inhabitants of the sea. A perfectly preserved skeleton of a female *Stenopterygius*, found in the Jurassic sediments of Holzmaden in Germany, shows that she died in childbirth, because inside her skeleton were found the embryos of unborn young, and all around her were the bodies of tiny ichthyosauri representing the individuals already born. This discovery is extremely interesting since it proves that these marine reptiles had already reached such a degree of adaptation to aquatic life, particularly in open waters, that they no longer laid eggs, but gave birth to live young: an unusual example of ovoviviparity in reptiles. Easier to identify and decidedly more useful to the palaeontologist's purposes are the cases of death due to external factors, the instances of violent death. These cases stem from a great variety of causes: various kinds of accident, environmental changes, local catastrophes, attacks by other organisms, forays into hostile habitats, and natural traps. The latter can be found in many places — for example, the pools and lakes which dot the African savannas and trap in their often soft mud the larger mammals driven to them by thirst. Such was the landscape of the Gobi desert in the past. Its Oligocene sediments have yielded a splendid specimen of *Baluchitherium*, a relative of the rhinoceros over five metres tall, found in an extremely unnatural position for a fossil. The huge beast had been swallowed up by the sediments while standing on its four legs, a certain sign that it had suddenly sunk in the mud and had been unable to free itself.

The best example of a natural trap is the asphalt lake of Rancho la Brea, near Los Angeles, California, in the United States of America, where Pliocene mineral oils reached the surface through Quaternary soils and solidified in irregular patterns. It often happened that an animal approached these asphalt surfaces, either looking for water or running from enemies, and was trapped by the fluid mass, which quickly submerged it and preserved it for posterity for hundreds of thousands of years. These asphalt strata formed during the Quaternary have yielded the best collection of vertebrates handed down to us from this period. They included, among many others, specimens of *Smilodon*, the sabre-toothed

An example of perfect fossilization: a crustacean of the species Eryon arctiformis *found in the Upper Jurassic deposits of Solnhofen, Germany. The Solnhofen site is famous for the perfect preservation of its fossils, which are found on thin slabs of yellow limestone. Below: two more crustaceans typical of the site*

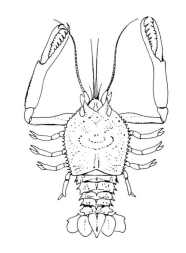

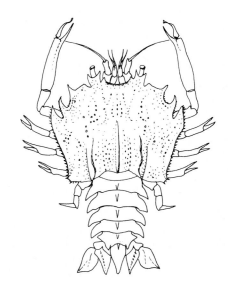

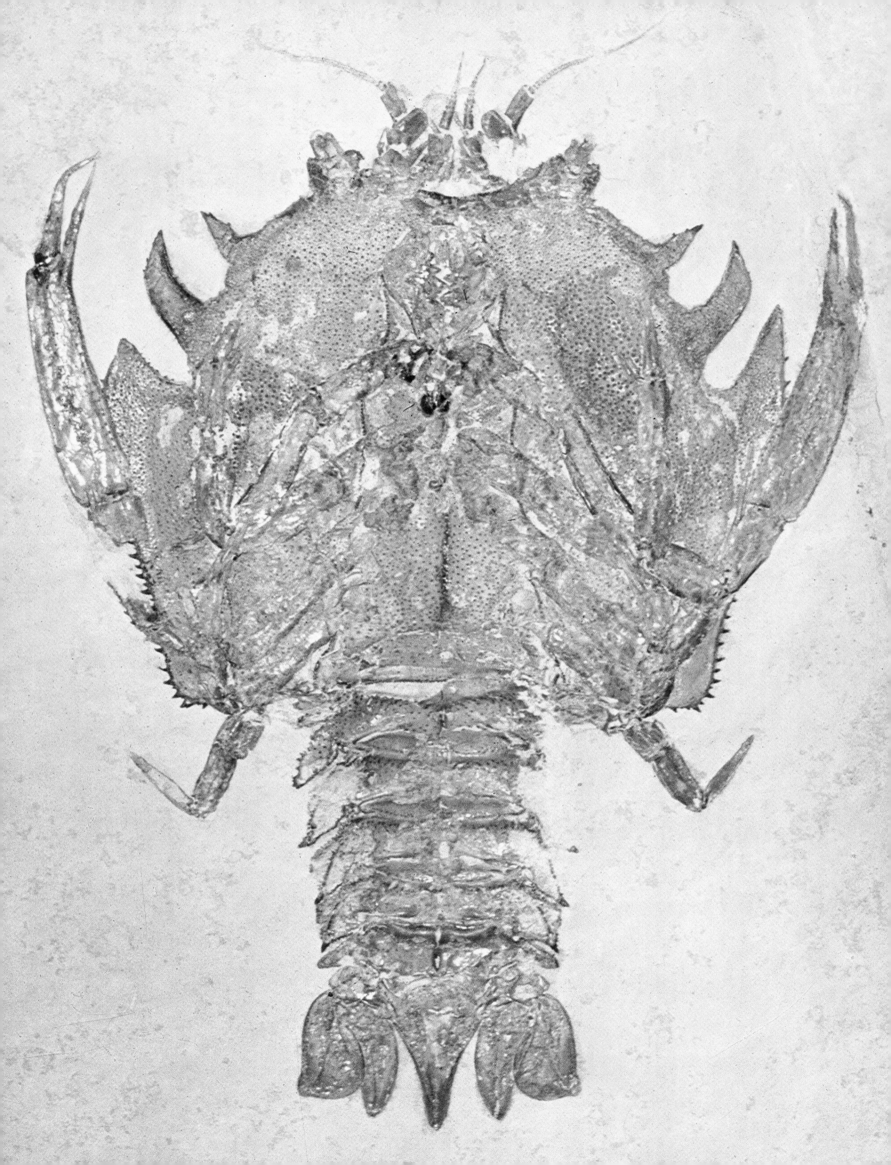

tiger, and *Archidiskodon imperator*, the gigantic elephant of North America. What is extraordinary is that this process of trapping and fossilizing still happens: it is therefore possible to observe "in the round" all the phases which led to the preservation of extinct organisms.

Another famous natural trap is the one which captured the Beresovka mammoth in Siberia where these cold-climate elephants had made their last stand. A huge specimen of a mammoth was found here enclosed in ice and still holding in its teeth the remains of the plants which constituted its last meal. The creature had fallen, twenty-five thousand years ago, into one of the many crevasses which crept across the frozen surface, had been unable to free itself on account of the many fractured bones caused by its fall, and had been frozen within the permafrost layer.

Another source of violent death is the fight for survival. All organisms in the wild belong to a chain in which the presence of each is influenced by the presence of the others. Thus plants provide food for herbivores, and the latter enable carnivores to survive, and so on. This relationship between prey and predator is one of the commonest causes of death; palaeontologists know of several examples of it. The shell of an ammonite of the genus *Placenticeras*, found in Cretaceous rocks of the United States, shows signs of biting caused by a marine vertebrate of the genus *Mosasaurus*, a gigantic lizard living in the same geological period. It looks therefore as if ammonites were part of a mosasaurs' favourite diet. The stomachs of other Mesozoic marine reptiles have yielded belemnites, ammonites and fragments of other invertebrates which were all part of the diet of these rulers of the seas.

Yet another cause of death can be found in the variations of environmental conditions. Such a cause, when established, is particularly interesting in the context of a reconstruction of geological events. A clear example of this can be found in the corals which built the huge reefs: they need special conditions, certain degrees of salinity, temperature and depth, and the slightest variation in their habitat causes them to cease all activity and die quickly.

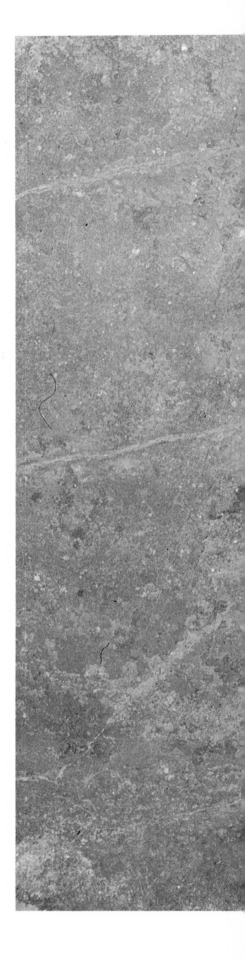

Fossilized skeletons are often found with their bones separated, as a result of those occasions when the fossilization process only commenced after the decomposition of the soft parts. This is what happened in the case of this marine reptile, Pachypleurosaurus edwardsi, *found at the site of Monte San Giorgio, canton Ticino (Switzerland): the two mandible bones have separated and detached themselves from the skull*

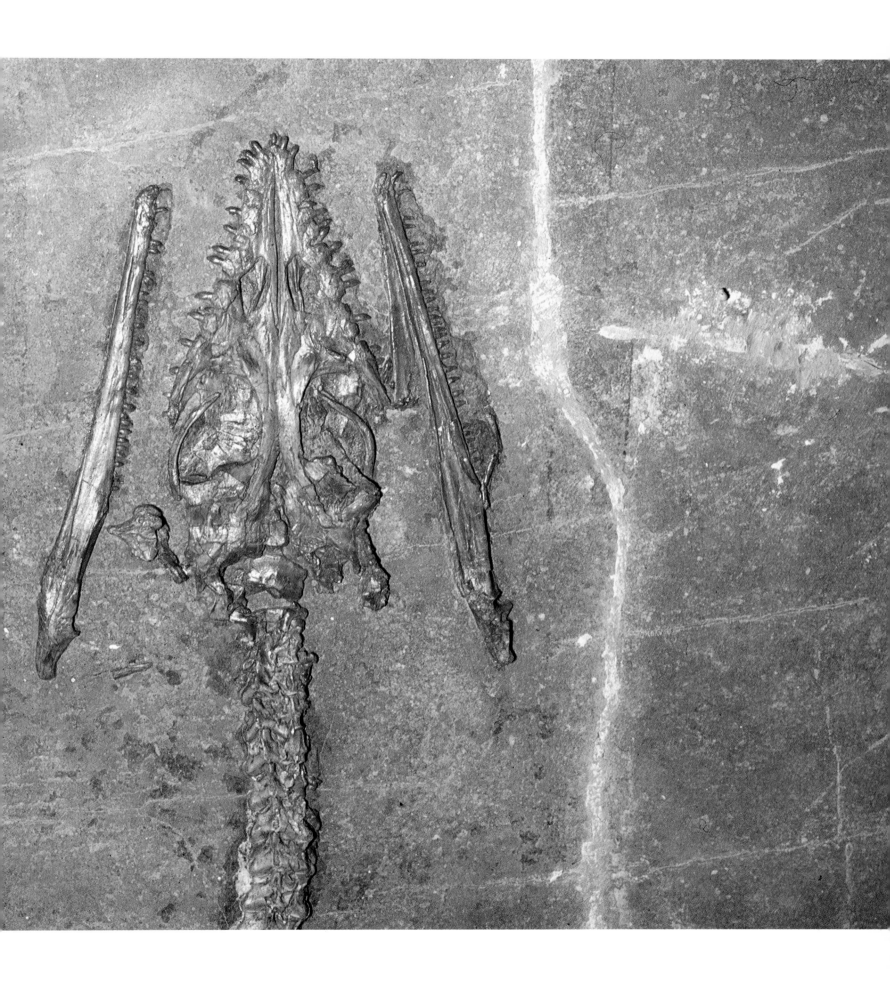

Finally, death can be caused by an organism venturing into a hostile environment. Classic examples of this are the sediments of Holzmaden and Solnhofen, in Germany. During the Jurassic, these two localities appear to have been coastal lagoons separated from the open sea by various types of barrier. They were closed-in environments, probably over-salty and not particularly suited to life. Violent waves could throw several kinds of organisms over the barriers into the lagoons, where they would find unsuitable conditions and would almost certainly and rapidly die. These two localities offered conditions not only unsuitable for life but also particularly well-suited to the preservation of the dead organisms: plentiful fine sediments, lack of currents, absence of mechanical or bacteriological destructive agents and of other biological factors. It is therefore easy to understand both the quantity of fossils found in these rocks and their high degree of conservation. In the Solnhofen deposits, scientists have found, by the side of fossil king-crabs, the traces of the spasmodic movements the animals made in their agony, an obvious sign of death due to unfavourable environmental conditions.

A lot less common than one might think is death due to local and natural catastrophes, such as sudden volcanic eruptions, flooding, etc. An example of one such death, caused by a local catastrophe, can probably be found in the famous Eocene deposits of Monte Bolca, near Verona, in Italy, one of the largest fossil deposits in the world. In its calcareous strata can be found all sorts of organisms: invertebrates, plants and many species of fish almost perfectly preserved. During the Eocene, the Bolca area was a coastline with many beautiful bays and coral reefs, surrounding lagoons full of plants and animals. The whole region was, however, affected by violent volcanic eruptions, as testified by the pyroclastic rocks which still come to the surface here and there. The sudden increase in water temperature and the poisonous gases produced by the eruptions killed large numbers of fish and other forms of life. The warm ascending currents then carried these remains to the surface and deposited them inside the small enclosed lagoons, one of which has been preserved in the form of the Bolca sediment. The water was calm here, and the rapid sedimentation allowed perfect preservation of the organisms.

When fossilized skeletons still retain their bones in the original articulation, it becomes much easier to reconstruct the organisms. This flying reptile of the genus Pterodactylus *was found at Solnhofen (Germany). The bones, still in their original position, show the wing structure which was supported by the elongated bones of the fourth finger*

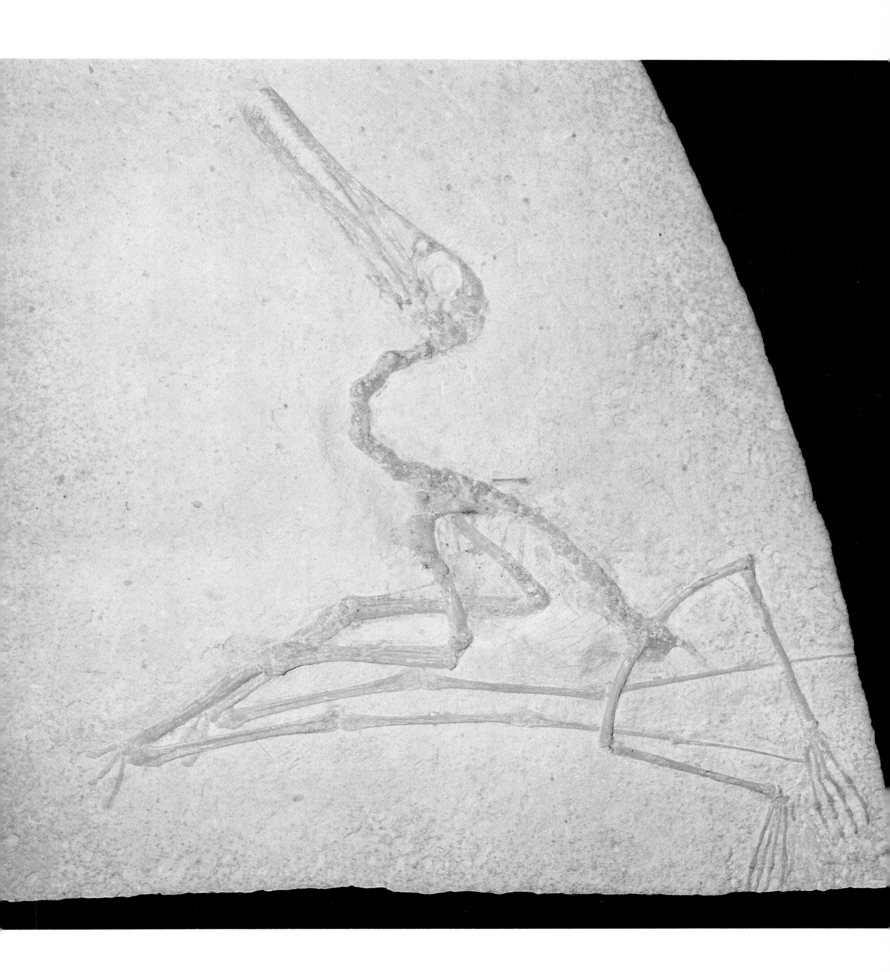

The Fossilization Processes

A coral cup turned to silica illustrating the process of fossilization through mineralization. Devonian of North America

The death of an organism is followed by complex chemical, physical and biological fossilization processes. They maintain the organisms in various degrees of preservation down to our own times.
The first stage in this fossilization process is one of removal, a so-called "post-mortem transportation". This is a very common phenomenon which has led to organisms fossilizing in areas different from those they inhabited, often considerably distant and characterized by environments thoroughly dissimilar from the original ones. Dead organisms are affected by many factors; sea currents can transport the lightest of shells for hundreds of miles from the place they lived in, while heavier creatures can be dragged for shorter distances. The degree of post-mortem transportation can therefore vary considerably.
When analyzing a group of fossils it is always extremely important to discover whether each of them is autochthonous or allochthonous — in other words, whether the fossil in question used to live in the place where it has been found or whether it has been transported there from somewhere else. When we examine sessile organisms (those which live attached to the sea-bed, such as corals) it is clear that in the majority of cases they were preserved in what was their original habitat. The transportation factor is more likely to have affected nektonic organisms which roam the seas freely, such as the ammonites. The latter floated in mid-water and, after death, must have been subjected at least to a vertical transportation, falling slowly to the bottom away from their original habitat. This is also the case with birds and the flying reptiles of the Mesozoic, the skeletons of which were preserved in a marine environment because they fell while flying over the vast expanses of water.
Death and post-mortem transportation are followed by a series of disintegration processes which act at different speeds on the various organic parts of the corpse. This disintegration can be due to three types of factors: biological, mechanical and chemical. The biological destructive agents which more frequently than others affect the soft parts of an organism (those which are more rapidly lost) include decomposing bacteria. Even

the most superficial layer of deposits contains bacteria and it is therefore essential, if the soft parts are to be preserved, for a rapid and thick sedimentation to take place and cover the organic remains before they are completely decomposed. Other organisms contributing to the destruction of the remains include boring sponges, scavengers, etc. It follows that the sooner the organic remains are protected from such destructive factors the higher the chances of their being preserved as a fossil.

The same principles apply to the mechanical destruction caused by currents, waves, wind and other factors which produce abrasion and corrosion of an organism and sometimes destroy it completely. The combined action of biological and mechanical destructive agents is such that, more often than not, fossils are found to be incomplete, broken and, above all, scattered over a more or less wide area.

Chemical change plays an important role in the preservation of organic remains, since it can continue to affect them even when they are already fossilized.

Chemical dissolution is more destructive to the organism's soft parts, constituted of carbohydrates and proteins, and has less effect on the hard parts, usually composed of calcium carbonate, calcium phosphate, silica or organic substances such as chitin and keratin which sometimes form highly resistant structures. Among the latter one would find the shells of molluscs, the calcareous skeletons of corals and the chitinous carapaces of arthropods.

The resistance of these hard parts varies within any given group and can depend on the degree of calcification which had been reached by the time the animal died. Even within the same organism, certain parts are found to have fossilized and others to have vanished, since different structures react quite differently to chemical corrosion. In mammals, for instance, teeth are more easily preserved than bones; they are less resistant to dissolution.

Finally, a lot depends on the type of sediment which has surrounded the organism. It has been found, for instance, that the same species of mollusc is preserved in a completely different way when buried in clay rather than sandstone. Generally speaking, one can state that coarse sediments — such as sand, conglomerates and gravels, easily infiltrated by waters — do not lead to a good preservation of organisms, so that fossils are rarely found in such rocks. On the other hand, clays, marls and all impermeable sediments are much better suited to give the necessary protection and are therefore found to be rich in fossils.

It could be said that the best deposits of fossils are found in finely grained sediments formed within aquatic environments in which the main agents of mechanical, chemical and biological destruction were only sparsely represented. It must by now be obvious that the environment in which sedimentation took place plays a major role in the preservation of organisms. A marine environment, where sedimentation is more abundant, more even, and formed of less coarse elements, is more propitious than a continental environment where poor sedimentation and the more violent action of destructive agents, such as wind, tend to preserve only part of the organisms. There are, of course, exceptions: in a marine environment, the rocky areas near the coast are less suited to fossilization

Fragments of Carboniferous ferns fossilized in talc minerals. Carboniferous of North America

On the following pages: *the large photograph shows an ammonite of the English Jurassic mineralized in pyrite.* Top left: *a silicified gastropod found in India.* Below: *a brachiopod of the genus* Paraspirifer *in pyrite of North American Devonian.* Right: *a fossilized Australian echinoderm transformed into azurite*

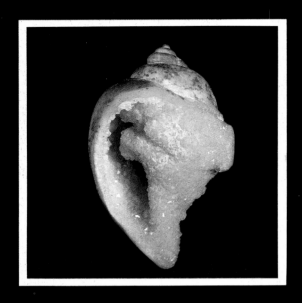

as the waves produce constant movement and sedimentation is coarser. A continental environment, on the other hand, offers good fossilization conditions in lakes, swamps and loess deposits where sedimentation is particularly calm and abundant.

A continental environment has a further effect on the preservation of organisms: under certain conditions, concentrations of fossils can form over a long period. While making research easier, this can also cause problems for scientists wishing to reconstruct the original habitat, as these accumulations can contain organisms originating from the most varied environments.

The waters, rich in dissolved mineral salts, which circulate within sediments, affect the organisms embedded in the sediments in two different ways: they tend to dissolve organic remains while at the same time they impregnate the organism with mineral substances, thus stabilizing it and preserving it. The latter process is usually called mineralization.

In the simplest of cases, this substitution of organic matter with inorganic mineral substances is total: the hollows which, in the organism, contained organic matter are filled by the mineral substances and the organism gains in strength.

Molecular substitution or diagenesis is a rather more interesting process: it consists of the substitution of each and every organic molecule. A complete modification of the chemical composition of the organism is thus produced, but its shape remains unchanged and is preserved down to its most minute details. Some examples of this process are fossilized wood which still shows the growth rings, the shells of foraminifers with their chambers still intact, the shells of molluscs which often preserve the various layers they are made of. The most frequent mineral agents in the mineralization process are undoubtedly calcium carbonate and silica, closely followed by calcium phosphates, pyrite (in those environments with a high percentage of organic substances), lead or zinc phosphates and many sulphates.

Calcareous or siliceous fossils are extremely frequent: whole forests have been preserved by the trees being transformed into opal or chalcedony blocks often retaining the original position with their roots still in the ground. It is also worth mentioning the possibility, however rare, of organisms fossilizing within minerals, such as silver, which produces real natural jewels.

Carbonification is a fossilization process which affects mainly plants and has led to the vast seams of coal formed during the Carboniferous, at least 340 million years ago. During this geological period, large areas corresponding to modern China, India, Australia, Africa, North America and parts of Europe, were covered by huge swamps surrounded by lush forests, the growth of which was encouraged by a warm and humid, tropical climate. The remains of these ancient forests form the basis of coal deposits.

It is not known exactly how these deposits formed; it is however believed that they formed on the very location of the forests, a theory supported by the discovery, within the deposits, of some plants embedded in their original growth position with their roots still in the soil. Other authorities believe the huge accumulations of plant material in coal-seams to be due to

Amber provides rare examples of the fossilization of whole organisms. The resin of ancient conifers, it trickled down the trunks swallowing small animals, thus preserving them through the ages. Below and right: two fragments of Oligocene amber from the Baltic, containing respectively a beetle and an ant

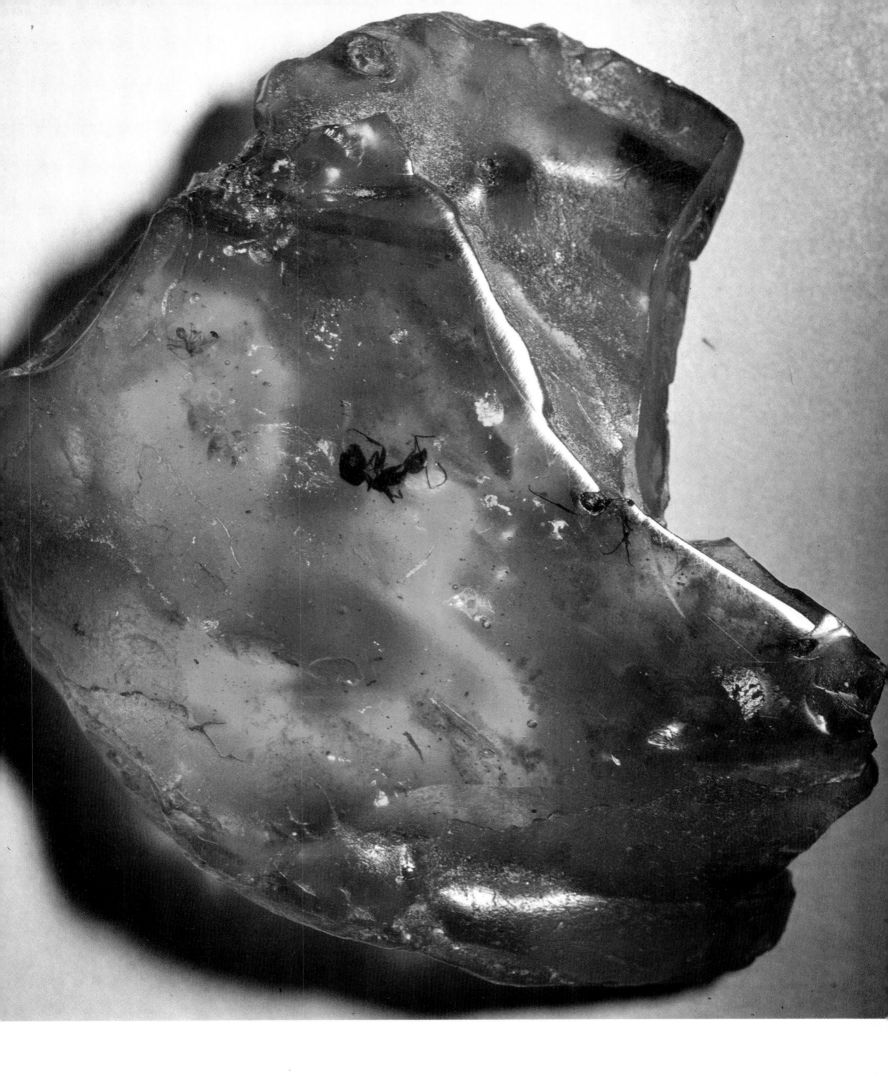

37

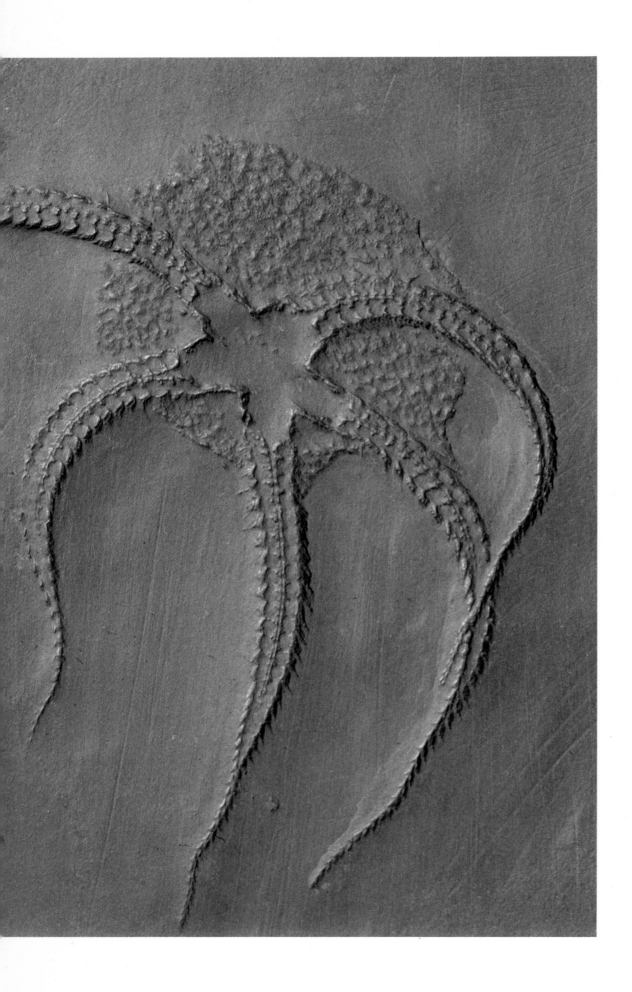

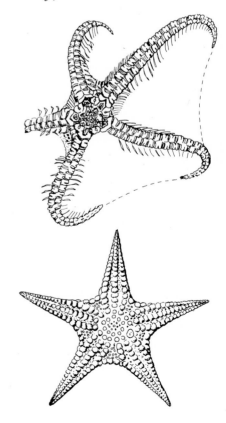

A starfish some 390 million years old transformed into pyrite: Taeniaster *of the Lower Devonian of Bundenbach (Germany)*

The wings of this dragonfly of the species Tarsophlebia eximia *still retain their delicate veins: a perfect example of the fossilization of perishable structures. Upper Jurassic of Solnhofen (Germany)*

the plants being transported, after death, to lagoons or coastal gulfs. Their theory is based on the fact that the coal layers also contain plants which must have lived in drier areas. It is very likely that both theories are founded on truth: coal deposits may well have been formed partly by plants living in situ and partly by plants removed from surrounding areas.

The carbonification process itself is due to the action of certain bacteria which attack the plant remains, eliminate oxygen and nitrogen and indirectly add carbon. The longer the period since the deposits were first formed, the richer the carbon content. These fossils are classified on the basis of their age, which means of their carbon percentage. Another interesting fossilization process is the one known as encrustation. It is limited to fairly recent fossils and is caused by waters rich in calcium carbonate which, flowing on the organisms, cover them in a thin coat of the mineral. Later, the organic remains are destroyed leaving only a negative print as a trace of their existence. A famous example of this is travertine — a rock containing a large amount of plant remains preserved

Leaves preserved in travertine constitute one of the classic examples of fossilization through encrustation. The specimen in the photograph was found in the Quaternary travertine of Tivoli (Italy)

The molecular substitution between the organic substances and the inorganic ones contained within the sediments allows the preservation of the smallest anatomical details of the organisms. The photograph shows, in section, part of a graptolite colony of the Silurian; all the tiny cells are perfectly preserved

in this way; the Ancient Romans used it widely to decorate their own homes.

A further fossilization process, however rare, is the one known as distillation, whereby the most volatile elements among those which make up the organic remains are distilled and leave a thin film of carbon on the rock, preserving the original shape of the organism. Fossils thus produced are far from perfect: graptolites, for instance, which more than other organisms were preserved in this way, were only fully understood after the discovery of some specimens preserved through a process of mineralization.

Under certain conditions, or due to an unusual combination of factors, some organisms have been preserved in very unusual ways: they still possess not only the tougher structures, as is the rule, but also traces of those soft parts which normally disappear completely. One of the deposits where conditions permitted the preservation of these extremely delicate structures is the famous one at Solnhofen, in Bavaria (Germany). Here,

Organisms are not always preserved with their external shape intact; often only their impressions are left on the rock, or the cavities formed by the decomposition of the body. Such is the case of this gastropod, lined with calcite crystals, which was found in the Jurassic limestones of the Furlo pass (Italy)

the skeleton of the earliest known bird, *Archaeopteryx*, was found and its identification as a bird was made possible by the exceptionally good preservation of its feathers. The same deposit has yielded the remains of flying reptiles with traces of the wing membrane, tentacles of jelly-fish, insects complete with their thin and delicate wings, and belemnites with all their tentacles.

Such a high degree of preservation has been rendered possible at Solnhofen by the extremely fine sedimentation which quickly covered organisms and protected them from the action of destructive agents, as well as by the absence or scarcity of such destructive agents. Solnhofen lacked any intensive mechanical destruction as there were no strong currents; the area consisted of a closed basin, with very calm almost still waters which were probably not very well suited to life.

By the same token, the Burgess shales in British Columbia (Canada), believed to date from the Cambrian, and those in Holzmaden, Germany, dating from the Jurassic, have preserved some exceptional specimens: highly delicate organisms perfectly preserved thanks to the absence, in the ancient environments, of biological or mechanical destructive factors. Holzmaden was a marine environment with almost still waters, a kind of lagoon, with no oxygen and therefore with no bacterial life, a stretch of sea with neither waves nor currents.

Another example of how extremely delicate organisms can be perfectly preserved is represented by amber. This resin, trickling down the trunks of conifers, swallowed up the various insects and spiders which lived in the ancient forests, protected them from external agents and later polymerised, preserving them intact to the present day. The best deposits of amber are the Oligocene ones of the Baltic, Romania, Sicily and the Apennines.

A last process of fossilization, of which only a few examples are known, is the one known as mummification, which involves the total conservation of even the most delicate parts. Two specimens of *Anatosaurus*, a dinosaur of the Cretaceous, have been found complete with their wrinkled skin stuck to the bones, as if the animals had undergone a thorough dehydration after death. Scientists believe this might have happened because the two specimens were buried by sand and therefore insulated from the effects of sedimentary waters; furthermore, the mineral contents of the sand would have caused the almost perfect petrification of the skin.

As we have seen, the waters which percolate through mud often cause the destruction of the organic remains, or else they impregnate them with mineral substances. In the first instance, should the sediment be well advanced in the process of solidification, the demise of the organic remains leaves a hollow mould which is soon filled by other substances. This produces a cast which reflects the external appearance of the organism and is called a pseudomorph. When, on the other hand, the organism is impregnated by mineral substances, the resulting fossil is not a cast of the organism but the organism itself, however transformed.

An internal cast is produced when the interior of a shell or of any other organism has been filled by mineral matter prior to the dissolution of the external layers or body. Internal casts are quite common but difficult to classify.

Wood is usually very well preserved as a fossil: the silicified palm trunks of the Sahara and those of the conifers of Arizona's petrified forests are famous. The photograph shows the section of a silicified trunk with the ancient wood structure

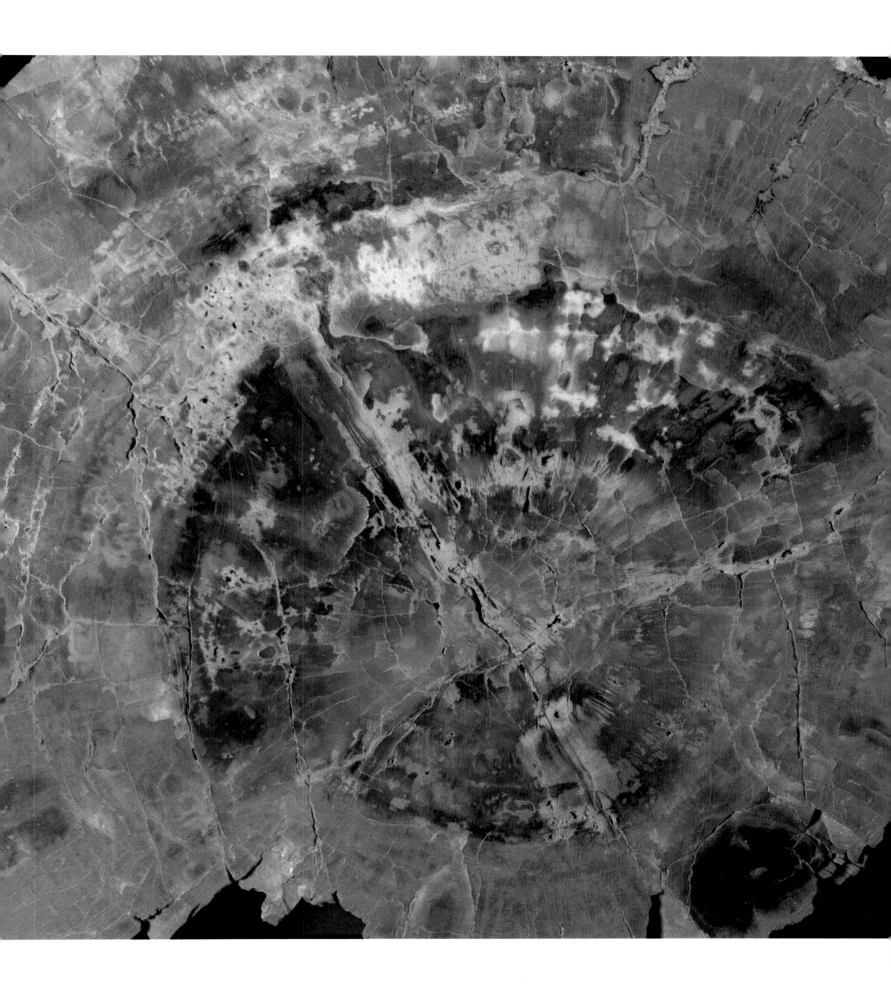

Rocks

Marine limestone containing numerous fossilized remains of the brachiopod Rhynchonella. *Carboniferous of Great Britain*

In the preceding chapters we have dealt with the various processes which affect organic remains before they turn to fossils, or during such transformation, and we have seen the close relationship between the organism and the sediment which imprisoned it. It can often be said that the sediment and the fossils contained within it were formed at the same time and in the same environmental conditions. However, this balance between time and habitat is often affected by external factors; the organic remains may have been transported, after death, to a place where environmental conditions were quite different from those in the place of origin, or geological phenomena may have interfered with a fossil after its formation and have redeposited it within older or younger sediments.

The study of the relationship between a fossil and the sediment around it has always been fascinating. Only by studying the surrounding sediment can anything be learned about the conditions of the habitat of a once-living organism which has now vanished and was quite different from any living ones. Our research is made easier if we observe what happens in present-day habitats, collect data on the modern relationship between habitat and sediment and, finally, compare these with the ancient sediments.

Classic examples of attempted reconstruction of an ancient living habitat are those which have affected the first vertebrates. These are mostly found as fragments in Ordovician rocks and can be identified as heterostracan agnathans — marine vertebrates without jaws or even fins — related to modern lampreys. The sandstones in which these fragments are embedded are typical of a marine coastal habitat, but this is not enough to prove that the ancient vertebrates were marine animals, as they do not appear in any other Ordovician marine deposits. It could be surmised that they might have been freshwater animals and had been transported to sea later on, when already reduced to fragments. However, there are no other contemporary continental sediments in the whole world which contain remains of vertebrates, so that the problem cannot be resolved with the palaeontological data so far available.

The first step in a serious palaeontological analysis is to define the relationship between fossils and sediments. In order to do this we have to clarify the meaning of sedimentation and sediment. Sedimentation is the result of an accumulation of material produced by the erosion of rocks. The process can take place in a continental environment, for example, producing dunes; or in a marine habitat. Whatever the location, we are always faced with layers of particles which vary in size and nature. The accumulation of these particles is called a sediment.

There are two main types of sediments: clastic sediments produced by transportation and sedimentation under the influence of such agents as water and wind; and chemical sediments produced by the precipitation of salts contained within the rocks. Just like the organic remains which, to become fossils, have to undergo certain chemical and physical processes, a sediment, once deposited, must be subjected to a change — called diagenesis — in order to become a sedimentary rock.

Sedimentation began with the Earth itself and the sediments since created make up the outer layer of the planet. These sediments form layers of varying thickness and depth which are closely connected with the time factor and which contain fossils; the history of the Earth is written within these layers. Successive layers of rocks are called (by palaeontologists and geologists) stratigraphic sequences. Stratigraphy is the science concerned with them. Thanks to the stratigraphic sequences we can analyse the various cycles of life and the succession of geological events, as long as the various layers can be dated.

The basic rule of stratigraphy states that any given layer is older than the layer immediately above it and younger than the one immediately below it. It is therefore quite clear that, since life has constantly changed during the successive geological eras, the study of the fossils contained within the layers of a stratigraphic sequence can give us a clue to the history of life on Earth.

The fossils contained within a sequence are therefore used mainly to obtain stratigraphic correlations and to date the rocks — the two main aims of stratigraphy. In other words, one has to establish which sediments were deposited in a certain area during a certain period and then relate these sediments to those deposited in other areas during the same age. All this can be achieved with the help of fossils. On the basis of various associations between the flora and fauna they contain, the stratigraphic sequences have been divided into a number of time brackets, easily recognizable, and into lithologically different rocks. Such divisions are made possible by the presence of the so-called zone-fossils, i.e. animal or plant species with a wide geographical distribution and a restricted temporal one. Ammonites are an excellent example of this type of fossil: such is their evolution that many species of ammonite lived for a relatively short time but over such a wide geographical area that they can be found in the rocks of stratigraphic sequences which are often very far apart. When the same species of ammonite is found in different layers and different areas, one can conclude that those layers were deposited in the same time bracket.

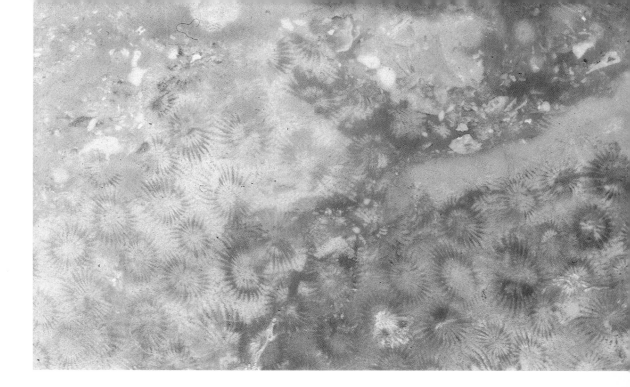

Section of marine limestone clearly showing the cups of the individuals forming a coral colony. Tertiary of the Apulias (Italy)

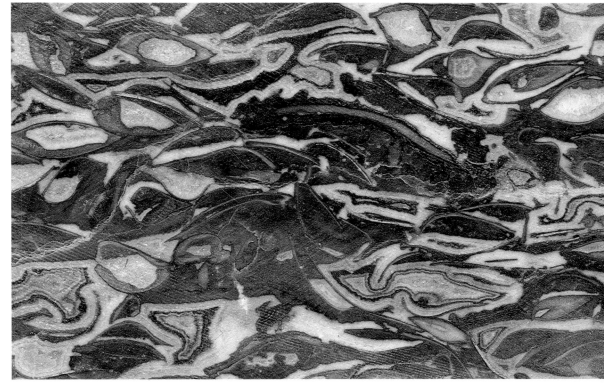

Section of organogenic limestone formed by an accumulation of molluscs' shells. Triassic of the Brembano Valley, Bergamo (Italy)

Surface of sandstone layer bearing the impressions of the inner sides of small lamellibranchs. Lower Triassic of Passo Rolle (Italy)

Dating the Rocks

From the palaeontologist's point of view, one of the more interesting results of the study of fossils is the possibility of a relative dating of rocks; that is, of finding out whether any given layer is greatly or only slightly older or younger than those above or below it. To achieve this one has to remember that the fossils contained in the sediments represent the evolution of the organisms which have succeeded one another over the various geological periods. As usual, this dating system presents, in practice, some serious problems, first among them the possible presence of different, and consequently non-comparable, fossils within rocky layers produced by sedimentation in widely differing habitats.

It is thanks to those zone-fossils with a wide geographical distribution that, over the past century and a half, a consistent dating of the Earth's rocks has been reached. It has therefore been possible to divide the history of this planet into periods which are independent from the type of rock one might encounter in the various stratigraphic sequences.

This type of division of the rock layers, called chronostratigraphic, has an absolute, as opposed to relative, value because each interval corresponds to a period of time which has always elapsed whatever the conditions and the habitat. Quite another matter is a division of the rock sequences based on their lithological composition. The value of such a division is purely local because each kind of rock is closely linked to a precise habitat. Such a lithostratigraphic division, therefore, can only faithfully reflect what has happened in a restricted area. On the basis of the faunas and floras which have succeeded one another during the course of time, the history of the Earth has thus been divided into fundamental units, each characterized by forms of fauna and flora which will not be found in identical form in any other unit. The first subdivision is based on two aeons, the Cryptozoic (i.e., "of hidden life") and the Phanerozoic. The former includes the Archaeozoic era, during which life was non-existent or very scarce, the latter the Palaeozoic, Mesozoic, Cenozoic and Neozoic eras, during which organisms developed and differentiated. Each era is, in turn, divided into shorter intervals of time, and these into shorter ones still, called periods, epochs and ages.

Each of these divisions has been called after the areas where the rocks belonging to it were first found, or where they were more abundant. Alternatively, they have been called after ancient populations who lived in those given areas. The Jurassic period, for instance, is called after the Jura mountains, which are made of rocks going back to that period; the Silurian period is named after the ancient Silures, who inhabited Wales; and so on. Unfortunately, palaeontology is helpless when it comes to define the length and exact age of the various chronological divisions, or rather their absolute age in number of years. The problem of defining the absolute age of rocks has always fascinated scientists and has only recently been solved thanks to modern radiometric methods which rely on the presence of radioactive elements in some types of rock or fossil.

This method is simple enough: once a molten mass consolidates, each radioactive element emits a series of radiations with constant speed and

Organogenic rock called lumachella, formed by the accumulation of small ammonites of the genus Arnioceras. *Lower Jurassic of Lyme Regis, England*

Fragment of a Miocene marl of the Apennines containing the fossilized remains of several lamellibranchs of the genus Cardium

slowly becomes a stable element. If one knows the transformation time of a given radioactive element and if it is possible to make a percentage calculation between the still unchanged radioactive matter and the stable element which derives from it and can be found in the rocks, it is then quite simple to deduce the age of the rock itself as well as the time of its consolidation. Uranium, for instance, changes in the course of time into various elements ending up as lead — the final and stable stage of the process. This transformation takes 4,560 million years; and on this basis it has been possible to date extremely ancient rocks. Uranium has been the determining factor in dating the oldest of rocks, those going back to about four and a half billion years ago. Other radioactive elements, such as potassium, sodium and argon, have enabled us to date many magmatic rocks as well as the fossiliferous rocks which are usually found with them. As a result, the dates usually quoted for any given geological period or event are not the result of simple deduction, but correspond with an exact reality.

Gastropods of the genus Cerithium *preserved in a Tertiary rock and indicative of a brackish sedimentary habitat*

Correlations

Stratigraphic sequences, as we have seen, are closely linked to the habitat which existed at the time and in the place of their formation. Environmental changes conditioned the type of sedimentation, its speed and the quality of the sediment itself, and caused the formation of layers of different rocks in different localities. No sequence exists which represents, with any degree of continuity, the whole history of the Earth. In order to reconstruct such an ideal sequence and to know all the biological events which took place in the course of the geological eras, it is necessary to correlate the various sequences, or rather the various layers of a sequence. By such correlations we can attempt to solve two major geological problems: the reconstruction of a continuous temporal succession by filling in the gaps usually present in stratigraphic sequences, and the identification of rocks of the same age deposited in different places and under different environmental conditions.

The problem of reconstructing a continuous temporal succession is not easy to solve: erosion, everywhere active, has often removed parts of the sediments, tectonic phenomena have folded, broken and often moved whole stratigraphic sequences, so that it is impossible to state with any certainty that they are now sited on the exact place where they were once deposited.

One of the most frequent gaps a palaeontologist can meet with is the so-called "syn-sedimentary gap". During a given sedimentation process, for instance in a marine environment, underwater currents can easily remove part of an existing sediment; the fossils contained in these layers will then be either destroyed or redeposited in underlying sediments of a different age. They are said to have been "reworked". Gaps of this kind give rise to enormous problems, particularly since they are extremely difficult to identify. A palaeontologist must always guard against the pitfall of identifying as contemporary certain fossils which, because of this or other geological events, only *happen* to be found together and which lived in different periods. Such interruptions in the sequences are very common; even when one finds a sequence of layers which has been deposited without interruptions, there is always the possibility of the fossils belonging to different ages and having come together by chance. These mixed deposits of differently dated organisms are often due to palaeontological condensation, another cause of headaches for the scientist and a very common phenomenon. As we know, sedimentation can vary enormously in intensity, and can even be totally absent. However, a standstill in sedimentation, during which no material is deposited, does not correspond to a standstill in life. It follows that the organisms which lived during this period of lack of sedimentation will get mixed up with the remains of organisms which lived before them, thus causing one of these mixed deposits which are so difficult to sort out.

One could also add those cases where a stratigraphic sequence has partially surfaced, thus exposing whole layers and their fossils to erosion and consequent destruction; and the phenomenon of lithological variation, due to environmental variations, which can be experienced within a layer between one of its extremities and the other.

To sum up, correlations are rendered difficult by many complex problems. This has not prevented palaeontologists from reconstructing, as a complicated mosaic, an extensive picture of what has happened in the course of the geological eras. Scientists have been able to relate rocks belonging to the same age; they have found that a certain marine mud had settled at the same time as the deposition, on the continental lands, of a desert or lagoon-type silt; and they have established that while ammonites and belemnites lived in the seas, life on the land was dominated by dinosaurs; that while amphibians were conquering the dry lands, modern types of fish had still not appeared in the sea, and that when the first coral reefs were being built only a limited range of primitive scorpions, millipedes and insects lived on the land.

Oligocene rock formed by layers of various types of molluscs. This unusual rock is called "Sternberg cake" after the site where it was found

Palaeoecology

Palaeoecology is concerned with the reconstruction of past environments, in their physical and geographical characteristics as well as in their biological aspects. While an ecologist studies present environments and is thus presented with a full picture of organisms and habitats, a palaeoecologist has at his disposal only extremely fragmentary information, such as the remains of animals and plants preserved in the rock layers. By examining these fossils, he tries to discover what sort of relationships existed between the various organisms during their lifetime and what the environment was like in a given place at a given time. Palaeoecology, however, is not only concerned with the reconstruction of an ancient coastline on the basis of the presence of certain fossils; it also studies the fossils themselves to establish a likely relationship between prey and predator, as well as cases of symbiosis and parasitism.

The palaeoecologist has other aids at his disposal besides fossils; one of the most important of these is the sediment within which the fossils are contained. It is the sediment which faithfully reflects the environment which produced the deposit itself.

All environments, both present and past, consist of a series of physical, chemical and biological factors which characterize them and define them and which determine the distribution of organisms. Physical factors include the actual nature of the medium, such as water or air, temperature, light, depth and so on. Chemical factors include the salt content of water, and the amount of oxygen or carbon dioxide present in the medium. Biological factors include the mortality rate, competition with other organisms, etc.

There are four fundamental principles which determine the relationship between organisms and environment, and they are common to both ecology and palaeoecology. They are: the ability of organisms to adapt to the environment, the ability to adapt to a way of life, the environmental factors and the interdependence between the organisms themselves.

Cyphosoma koenigi, *a sea urchin of Germany's Upper Cretaceous. The discovery of this type of fossil is read as indicating the existence, in the past, of a rocky marine coastline*

This conglomerate of fossilized organisms shows which were the principal life-forms on Silurian reefs about 420 million years ago

1 Posterior part of a trilobite

2 Shell of a brachiopod of the genus Spirifer

3 Fragment of a tabulate coral

An example of ancient symbiosis as documented by fossils: the annelid Hicetes *has built its tubular home on the coral* Pleurodictyum problematicum *(Lower Devonian, Germany)*

Flabellipecten flabelliformis, *typical lamellibranch of the Pliocene, Central Apennines*

1. Each organism has adapted to a particular environment which could vary considerably in latitude. Thus, while ammonites lived in all seas with no distinctions of salinity or temperature, other molluscs could only live in more restricted and defined environments. The size of the area which a given organism can inhabit depends on the ability of the organism to adapt to more or less strong variations of the environment.

2. Each organism is adapted to a certain way of life. This means that an organism can react to environmental stimuli and modify its own body accordingly in order to survive. Thus, for instance, those vertebrates living in open waters and needing to be swift swimmers and efficient predators, have assumed a hydrodynamic shape. Different groups of animals react to the conditions of the same environment by assuming very similar shapes.

3. The third fundamental principle of palaeoecology is based on limiting

This tangled print was once thought to be the tracks left by a worm, which was named Lumbricaria. *It is now thought to be either the excrement of a fish, or the stomach of a holothurian (Upper Jurassic of Solnhofen)*

factors. It is these factors which, either singly or together, determine the distribution of organisms. In the case of certain corals, for instance, the determining factor for their survival is the constant temperature of the water, which must be always around 20°C.

4. The fourth principle is concerned with the dependence of each organism on the sum total of the other organisms living around it. The presence of certain predators in any given area excludes the presence of others while favouring the existence of organisms which can survive by feeding on the corpses left by the former. The result is a tight chain of inter-relationships. The slightest imbalance can have disastrous results and cause the disappearance of whole groups of animals or plants.
The first step towards the reconstruction of an environment is represented by the analysis of a sediment. Here one has to consider various factors: the type of rock, its mineralogical structure, its physical structure, and certain

Antrimpos noricus, *a crustacean found in large numbers in the Triassic deposits of Cene*

Archaeopalinurus levis, *the oldest known lobster, found in the Cene site (Italy)*

Protoclytiopis antiqua, *another crustacean from the Cene site. This deposit formed 200 million years ago in a closed marine basin. The water had a low oxygen content and was perfectly suited to the organisms' fossilization. The presence of crustaceans shows that the basin was surrounded by reefs teeming with life*

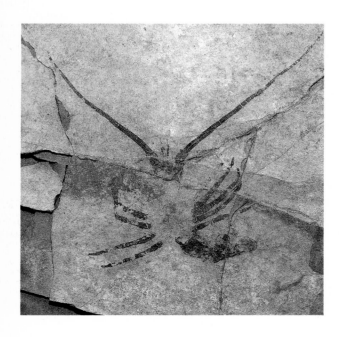

On the following page: *This Palaeozoic trilobite has fossilized in a coiled position, thus preserving the defence attitude typical of this group of primitive arthropods*

61

sedimentary formations sometimes found within the rocks. Clearly, any given type of rock can only form in a certain environment and not in any other. For example, a conglomerate points to a continental fluvial environment and cannot be attributed to a deep-water marine sedimentation. By the same token, the presence of evaporites like gypsum and rock-salt points to the fact that the original environment must have been a closed basin with strong evaporation. A bituminous rock, on the other hand, indicates a limited basin with low oxygenation and an acid environment on the bottom.

The stratification of a rock, the thickness of the layers, the presence of massive embankments also help us to reach a clearer definition of a particular environment.

Many sandstones, for instance, show a characteristic distribution of the granules, at an angle with the sedimentation surface. Such a distribution is called "crossed stratification" and indicates the prevailing direction of the currents at the time when the sedimentation occurred. There are other sedimentary structures: among them are the ripple marks or parallel wavy marks of the sea bed which are formed by the action of underwater currents on shallow beds; the glacial striations of certain continental regions, caused by past glaciers slowly dragging enclosed boulders over the rocks; the desiccation cracks of mudflats which clearly indicate the climatic conditions prevailing at the time of their formation.

When analysing a sediment, the palaeoecologist has to consider the often close relationship between sediment and organisms; this relationship can be both constructive and destructive from the point of view of scientific analysis. Various types of sediment, for instance, are formed by an accumulation of organic remains; among these are the siliceous rocks, the coral reefs, and the layers of coal. Often, however, the organisms themselves have actually destroyed those very traces and structures which would otherwise prove so useful in establishing the characteristics of an original environment. To give one example: mud-dwelling worms can cause such disturbance within a sediment by their burrowing as to obliterate the original stratification. However, any palaeoecological reconstruction of a given past environment must take into account an extremely important fact: it has to be remembered that not all fossils found in a rock can be actually used, because many of them happen to be there having been transported from a different environment after their death. A choice has to be made initially.

Biocoenosis and Thanatocoenosis

As we have seen, the difference between ecology and palaeoecology is considerable. While the ecologist is concerned with the flora and fauna living in a given environment — the assemblage of which is called "biocoenosis" — a palaeoecologist has to rely on a number of dead organisms such as fossils. The fossil complex discovered in a given locality and a given rock layer is called "thanatocoenosis". There is a lot of

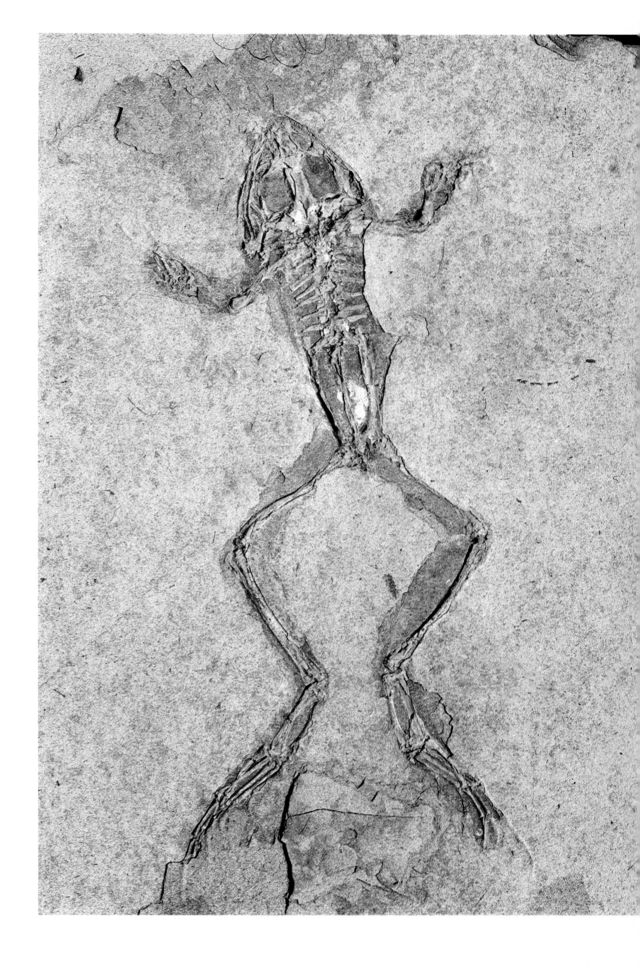

Rana pueyoi, *Teruel, Spain (Miocene)*. The existence of such a frog in a sediment indicates a lacustrine habitat. Frogs are rare as fossils; below is a drawing representing the oldest known frog, Triadobatrachus, *of Madagascar (Triassic)*

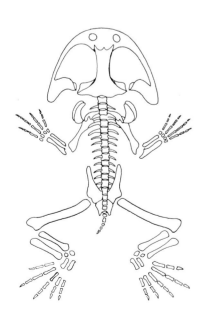

The branched tracks on this rock from the Apennines are called Chondrites *and were probably left by burrowing worms in a muddy marine bed*

difference between a biocoenosis and a thanatocoenosis. In the case of the former the scientist is confronted with organisms originating from the same environment presenting certain characteristics. The latter, on the other hand, can include various biocoenoses formed both by post-mortem transportation and by phenomena of palaeontological condensation.

Let us see what happens when the organisms forming a biocoenosis die. Some of them are preserved in situ (the so-called autochthonous elements), some do not fossilize at all for lack of suitable parts, and some are removed by various mechanical factors such as currents, wind and waves. This latter part will constitute the allochthonous elements of another thanatocoenosis. A scientist therefore has to differentiate, within a given thanatocoenosis, between the autochthonous elements, upon which an environmental definition can be based, and the allochthonous ones.

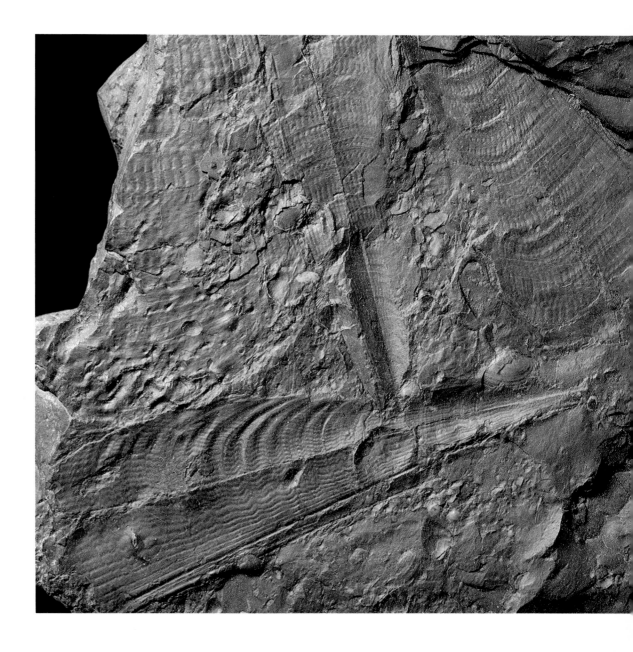

Shells of the genus Pinna *live today in shallow water mudflats. The same habitat must have applied to these specimens from the Triassic of the Bergamo pre-Alps (Italy)*

Clearly, sessile organisms, living as they do anchored to the substrate (e.g. certain brachiopods and lamellibranchs), stand a better chance of fossilizing in the same habitat in which they lived; the same applies to endobiotic elements living buried in the mud. Free-living organisms, on the other hand, are more likely to be transported and become allochthonous elements.

In order to ascertain with any certainty which are the autochthonous elements upon which our palaeoecological deductions can be based, we have to know first of all how the various ancient organisms lived. This is not easy, above all in the case of animals now extinct. However, other elements can help us in our difficult choice: if fossils are all oriented in the same direction, for instance, they are allochthonous, as their position indicates some form of mechanical transportation by current or other

means. The latter, moreover, tend to cause a certain selection as they deposit the fossil remains according to their dimensions, weight or shape. It is not rare to find accumulations of shells all belonging to the same species.
Only after having established which are the autochthonous elements of a thanatocoenosis can we begin our task of environmental reconstruction based on whatever information the organisms can provide.

Deductions from Fossil Organisms

Many fossils are similar, if not identical, to living organisms; it is therefore quite easy, in such cases, to establish their habits as well as habitats, which were probably not very different from those of their modern descendants. The seas of the Tertiary period, for instance, harboured many molluscs, lamellibranchs and gastropods, very similar to those living in modern seas; their habits must therefore have been the same. It is thus very easy to establish the original environment of such fossils, just as it is easy, when they are found in a given sediment, to deduce from their very presence the type of environment existing at the moment of their death.
Fossils which are characteristic of a certain environment and allow us to reconstruct the type of environment itself are called facies fossils. (Facies indicates the complex of organic and inorganic factors which determine whether a sediment belongs to a given environment.) Facies fossils are corals, crocodiles or freshwater fish which indicate precise life-styles; while ammonites, which lived in all sorts of environments and were widely distributed with no relationship to environmental characteristics, cannot be considered facies fossils.
It is not always possible, however, to establish so easily the type of environment in which organisms lived, particularly when we are confronted with groups now extinct. As if there were not already enough complications, it has been known for some organisms to change their habits in the course of time as they moved from one habitat into another. This is the case of *Aysheaia pedunculata*, the earliest onychophoran known, found in Cambrian marine sediments of British Columbia (Canada). It appears that this creature lived in the sea, while its modern descendants live on land, in the humid undergrowth of tropical forests.
But let us return to ecologically easier fossils. These ecological indicators, as palaeontologists often call them, help us to achieve environmental reconstructions which may seem incredible. They make it possible for us to calculate the depth of ancient seas or the climatic conditions of the various geological eras. Some fossils can actually be regarded as geological thermometers since they give us a good approximation of climatic conditions and temperatures prevailing in the place and at the time of their death.
Certain marine molluscs found in Quaternary deposits have been very useful indeed. Some belong to cold climates and some to warm climates and they follow one another in the sedimentary layers of this geological period, thus clearly indicating the climatic changes, the famous glacia-

It is very difficult to imagine the habitat of a given fossil when the latter is not comparable with modern organisms: this is what happens in the case of this Merycoidodon *of the South Dakota Oligocene, an early ruminant with teeth showing carnivorous characteristics*

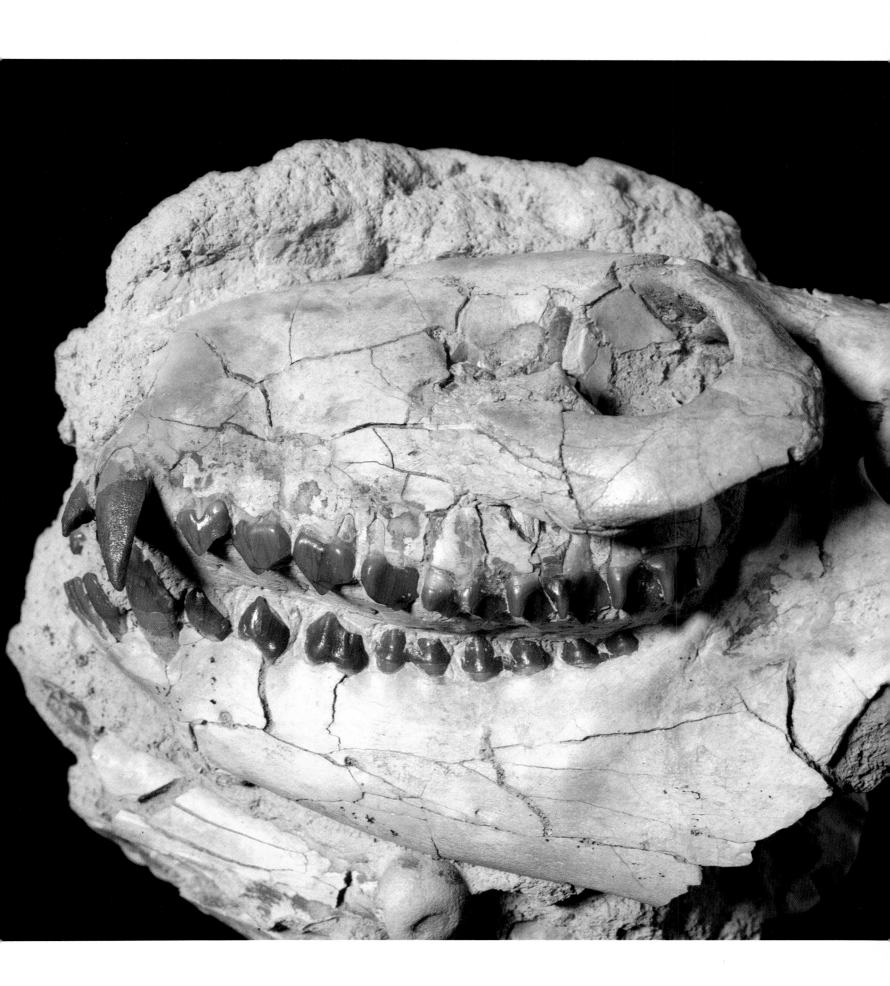

tions, which have occurred repeatedly during the last stage in the history of the Earth. The cold and warm guests, as they are called, would enter the Mediterranean during the glacial and interglacial periods, then withdraw southwards or northwards, via the Straits of Gibraltar, at the end of each climatic phase.

We have been able to reconstruct palaeoclimates even for extremely ancient periods. Triassic coral reefs, very common in the alpine area, indicate a warm, intertropical climate; coal-seams dating back to the Carboniferous have revealed that during that period the climate must have been very similar to the one existing today in some tropical areas. Palaeoclimatic reconstructions can be very difficult when we are faced with organisms which have no equivalent in modern nature; in such cases, a palaeontologist runs a serious risk of trespassing into the world of fantasy. Yet, even these organisms, however unknown, can supply us with interesting ecological information.

Let us, for instance, take the trilobites — a group of arthropods extinct since the end of the Palaeozoic. Some of their anatomical details are clear indications of a certain life-style. Some of them lacked eyes, which means that they probably lived buried in the sediments; others had a light exoskeleton, furnished with strong spines which probably equipped them for a floating pelagic life.

Even if a fossil organism provides no indication whatever to allow a palaeoecological reconstruction, all is not lost: other organisms may be associated with it in more or less close symbiotic or parasitic relationships, and they may lead us, by analogy, to discover the habitat of the mysterious creature.

Trace-fossils

Sediments and fossil organisms are not the only elements a palaeontologist can use to reconstruct ancient environments. Other such elements are the marks left by the organisms themselves while moving around and, generally speaking, all the traces of their biological activities. Clearly, these tracks and traces, which are preserved just like any other organic remains, are extremely useful in the exploration of ancient lives. The discovery of the footprints of reptiles and amphibians in late Palaeozoic rocks has shown, for instance, how these animals (of which we only know the skeleton) actually moved about. The footprints left by dinosaurs have enabled us to determine what sort of environment they lived in, how quickly they moved and whether they were gregarious or solitary. Sedimentary rocks are rich in such traces of biological activity; a trained eye can recognize them in such a rock and use them to reconstruct the environment in which the deposits were produced and even to suggest ecological relationships between the organisms in that environment.

However, footprints and trace-fossils are not always easy to identify. Although it may be fairly easy to attribute a footprint to a reptile, it is much more difficult to establish which kind of marine invertebrate might have produced a given trail or burrow. Indeed, there have been instances where the identification of the footprints of vertebrates has not been at all easy. The sediments of the European Triassic, for example, often contain very characteristic tracks, consisting of small and large footprints in pairs. Since the nineteenth century these have been attributed respectively to the fore and hind feet of an unknown type of reptile with larger hind limbs than fore limbs, which was named *Chirotherium*. Only in recent years has the discovery of skeletons of a carnivorous thecodont reptile, named *Ticinosuchus*, in the Triassic of Switzerland and Italy, provided us with an animal with limbs of suitable construction to have produced the *Chirotherium* prints.

In conclusion, it can be said that trace-fossils, such as footprints, trails, the galleries of burrowing organisms, fossilized excrement and the holes bored in shells by predators, are all valuable and informative elements in the reconstruction of the biological history of the Earth.

The presence of barnacles on this stone found in Pliocene sediments in the Apennines indicates the presence, in that period, of a coastline

Palaeogeography

This block of molluscs recently cemented together comes from the Venice lagoon and shows the first step towards the formation of an organogenic sedimentary rock. Rocks of this type help in reconstructing ancient coasts

Organisms of the past, like modern organisms, were characterized by well-defined geographical distributions frequently depending on those ecological factors such as the salinity, depth and temperature of water, which have been called limiting factors. Thus most animals and plants are bound to a certain habitat which results in their distribution being limited in time and space. If we examine the fauna living on a rocky sea-shore and on a sandy beach on the same coast, we immediately notice how different they are; but comparison of the organisms of two ecologically similar parts of the same coast reveals a corresponding similarity in the organisms, even if other different habitats intervene.

The limiting ecological factors are the cause of both the differences and the similarity.

When we examine the organisms of two environments which are similar but situated at great distance from each other, for instance on two different continents, we notice that they are no longer characterized by the kind of similarity which should be warranted by ecological factors. The two faunas are different even though they inhabit ecologically identical areas. The existence of such differences between the faunas and floras living in the various regions of the Earth is caused by geographical factors: those barriers, in other words, which prevent an organism from achieving total distribution.

In reality, there is no clear-cut distinction between ecological and geographical factors; what we call geographical factors are simply extreme ecological ones. A geographical barrier is only an ecological barrier which is so complete as to render it impossible for certain organisms to overcome it. Let us, for instance, take the Atlantic Ocean: its depth constitutes a geographical barrier which causes the shelf-fauna of the American coast to be different from that of the European shores. But this geographical barrier is purely an ecological barrier, insofar as it is its depth — an ecological factor — which prevents certain organisms from overcoming it.

There are various types of geographical barriers. Terrestrial organisms

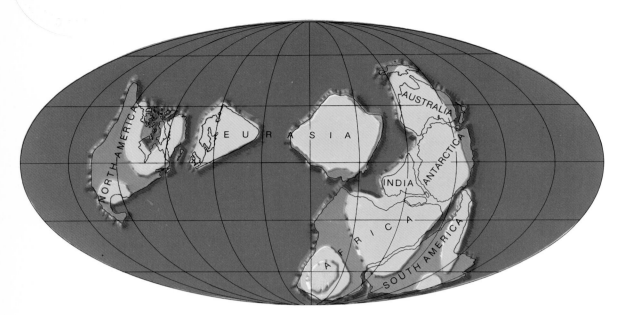

The theory of global tectonics has shown that continental masses have constantly shifted in the course of geological time, alternatively joining one another and separating again. The modern position of the continents is the result of such movement over a long period which has naturally affected the evolution of fauna and flora. The present distributions of organisms are the result of the movements of continental masses which, welding and separating the dry lands and forming mountain ranges, have influenced the evolution or extinction of organisms

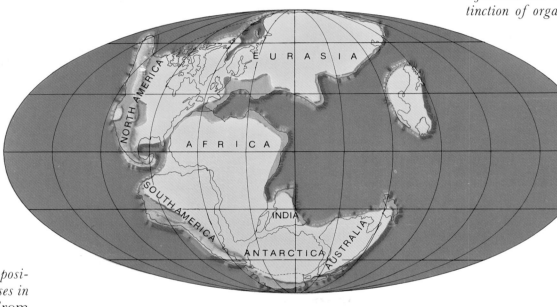

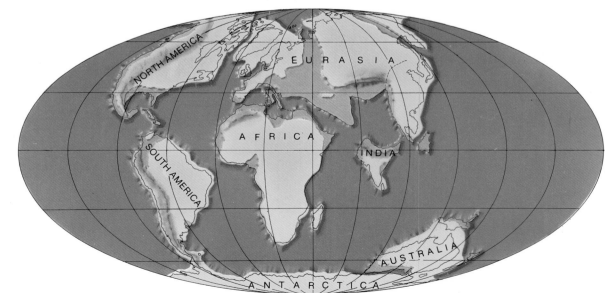

The three maps show the position of the continental masses in three geological periods. From top to bottom: *the situation in the Ordovician (500 million years ago) when the continents' position was completely different from the present one; in the Triassic (200 million years ago) when there was a single continent, Pangaea, with the great gulf of Tethys to the east; and in the Eocene (45 million years ago) when the continents' position approached the modern one. In the latter map, we can see that the Atlantic was still rather narrow, India still detached from Asia and Australia attached to Antarctica*

The distribution of fossilized plants and animals often shows quite clearly the shifting of continental masses. The discovery of identical animals and plants in localities which are today far apart, indicates that these localities were once connected. Several fossil organisms have led to these conclusions

Above: *a* Glossopteris *leaf, a pteridosperm which was common in South America during the Permian, as well as in South Africa, Antarctica, India, Australia and Madagascar. This type of distribution is the result of these continents being, at the time, joined into a single continent, called Gondwanaland by geologists*

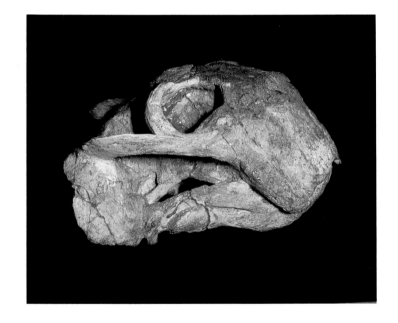

Left: *the skull of the continental reptile* Lystrosaurus, *which lived during the Lower Triassic; the animal has been found in South Africa, India, China, Russia and Antarctica, thus proving that these lands were once united*

Right: *the skeleton of the Permian aquatic reptile* Mesosaurus, *found both in South Africa and South America. Since* Mesosaurus *lived in fresh waters, and would therefore have been unable to cross wide stretches of sea water, its distribution indicates that, during the Permian, South Africa and South America must have formed a single continent*

may be restricted by mountain ranges, deserts, rivers and marshes; marine organisms by continents and great oceanic depths.

Another factor playing a very important role in the distribution of organisms is the time factor: the position of dry lands and seas has altered dramatically in the course of the various geological eras, and these changes have affected the fauna and flora of certain areas.

Modern South American mammals offer an excellent example of how geological events have affected the distribution of organisms. This particular fauna is the product of initial isolation of the groups which had settled on this continent during the lower Tertiary, followed by colonizations by other groups which in subsequent geological periods succeeded in overcoming the barrier caused by the submersion of the Panama isthmus, and finally by those forms which arrived from North America at the end of the Tertiary when the two continental masses were reunited by the rising of the isthmus.

The geographical distribution during the various geological eras has played a very important role in evolution; it has been demonstrated that the passing of time increases the differences between geographically separated groups, so much so that it can be stated that the longer two regions have been apart the greater the difference characterizing their faunas and floras. This clearly leads to a fragmentation of the distribution of organisms and facilitates that differentiating process which is called speciation and which leads to the formation of increasingly new and different groups.

Palaeontologists have been able to show that there must have existed, in the past as now, several fauna and flora provinces; in some cases, they have even been able to define their extent and characteristics. Jurassic ammonites, for instance, can be divided into three large provinces, a Boreal, a Mediterranean and an Arabian one. In each province, ammonites evolved differently, notwithstanding complicated relations of interdependence due to migrations of groups from one province to another. This brings us to palaeogeography — the reconstruction of the geographical characteristics of the Earth during the various geological eras.

All other sciences dealing with the Earth, from sedimentology to palaeontology, from stratigraphy to palaeobiogeography, contribute to this palaeontological branch. The latter is indeed one of the cornerstones of palaeogeography. Over the geological eras, the Earth's geography has constantly changed, each change modifying the preceding geographical features, so that there are no direct elements which may help us in determining such changes. Palaeontologists have to rely on the only certain data at their disposal: fossils and their successive distributions. In order to establish the geographical evolution of a certain area, therefore, we must base ourselves largely on fossils. As we have seen, the Panama isthmus has been subjected in turn to immersions and emersions, thus separating or joining North with South America. Such movements have been established thanks to the study of the fossils of marine molluscs which lived on the Pacific and Atlantic coasts of the isthmus. During the emersion periods, molluscs were different in the two localities, while

A young specimen of Mesosaurus tumidus. *Brasilian Permian*

during the immersions the two faunas were similar: a clear sign of a link between the Atlantic and the Pacific.

In the course of the Earth's history, the continents have continuously and gradually changed their relative positions. Between ninety and sixty million years ago, the various modern continents were produced by the splitting of a single large continental mass. This is the central concept of Wegener's famous theory of continental drift, which was partly based on palaeontological evidence, in particular the presence of very similar fossil faunas and floras on continents which are now widely separated.

Wegener's theory, put forward in the 1920s, was dismissed for lack of evidence for many years, but has become widely accepted in recent years, particularly as the result of the discovery of palaeomagnetism in certain types of rock. This has permitted the reconstruction of the past positions and movements of continents relative to the magnetic poles and thus permits the positions of the continents, and hence the seas, to be mapped for the various geological periods. Hundreds of scientists throughout the world are now exploring the consequences of these palaeogeographical data on their environmental and biological reconstructions.

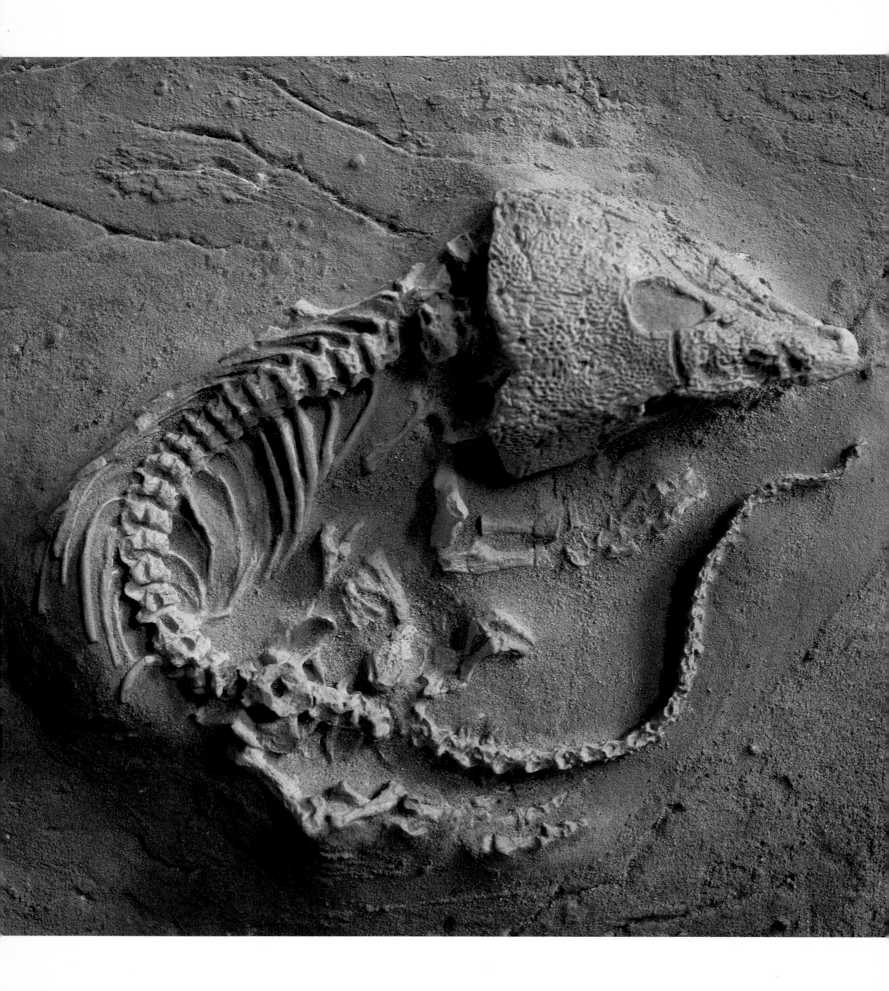

Evolution

The theory of evolution of Darwin and Wallace caused great controversy when first published, but has long since been accepted by virtually all scientists. Palaeontology has played a major role in this acceptance, since it is the study of extinct animals and plants which provides much of the evidence of the course of historical evolution. Continuing research on a steady stream of new discoveries through the twentieth century has given us a gradually improving picture of the historical evolution of organisms. It has sometimes been possible to demonstrate relationship between whole groups of animals or plants which had previously seemed to be systematically quite unrelated.

However, evidence of the rate and nature of evolutionary change is still very scarce in the fossil record. The evolutionary history of most groups is sufficiently fragmentary that there is little direct evidence as to whether evolution has been a gradual, continuous process or whether it has progressed in sudden spurts from one fundamental change to another. Some palaeontologists argue that when the fossil record is very good and can be studied at high resolution, then the fossils show gradual change from one species to another. This microevolution, as opposed to the macroevolution of the larger systematic categories, is still a controversial phenomenon and some palaeontologists believe the evidence for gradualism to be ambiguous. These controversies show that palaeontology is still contributing to evolutionary theory.

Palaeontology can support the theory of evolution with three kinds of evidence: whole series of fossils in continuous succession as documents of gradual variations within smaller systematic categories, usually one gradually changing species; fragmentary series of fossils which allow a broad reconstruction of the history of the larger systematic categories; and isolated fossils which, being transitional between different groups of organisms, are called link fossils.

Microevolution

Certain basic rules must be followed in order to trace the minor evolutionary transformations within a sequence of fossiliferous layers. Such rules are indispensable to the reconstruction of the history of a group of

Labidosaurus hamatus, *Texas Permian. This animal, not much longer than half a metre, is one of the most primitive reptiles we know. It belongs to the group of the captorhinomorphs, which we believe may have given rise to the majority of known reptiles*

organisms. It is first of all necessary for the layers under examination not to be affected by geological phenomena, to retain their normal succession, without any gaps and to show no trace of condensation phenomena. Fossils have then to be collected with great care, layer by layer, so that the provenance of each specimen can be carefully reconstructed, in the laboratory, taking into account its original place in the succession.

Complete and continuous sequences are unfortunately very rare and correlations are therefore often necessary to obtain larger series or to fill the gaps which may be present. Correlations carried out over large areas also help to avoid those errors which may be incurred when only local series are studied. They provide a more general view of mutations caused by the environment which may have taken place locally and have nothing to do with the evolution of the group in question; or, on the contrary, they may stress those phenomena of migration and geographical variation which play such an important role in evolution.

Among the invertebrates, the ammonites have supplied the best examples of continuous evolution. The study of these extinct molluscs has shown that their evolution could take place through successive modifications within a given group without new species branching off, or else through speciation phenomena — through a species subdividing into two or more groups for geographical or ecological reasons.

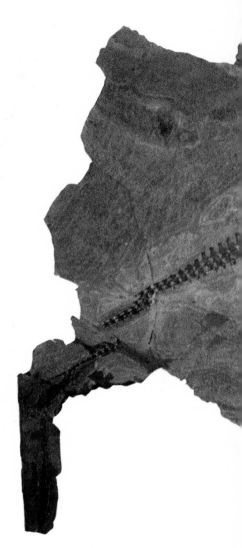

Macroevolution

Macroevolution is the evolution of the large systematical categories; not so much the transition from one species to another but the formation of new animal and plant types. However, since nature's fundamental unit is the species, the evolutionary transition between the larger groups boils down to a progression from species to species which takes place as a continuum. Macroevolution and microevolution are therefore the same evolutionary process and the formation of the large systematical categories must have happened through gradual and continuous variations. The lack of continuity which may appear to characterize the origin of the basic types is almost certainly due either to our incomplete knowledge, or to certain evolutionary processes which cause apparent discontinuities between the adult specimens of successive groups descending from one another. Such discontinuities are not real, however, and soon vanish when the specimens of successive groups which are being compared are taken at a stage of development other than the adult one.

Let us see how the origin of a large systematic category actually corresponds to the origin of a species. Let us take as an instance the horse family, the history of which began in the Eocene, some fifty-five million years ago, and is still going on. As with all the large systematic categories, the assignment of the family level to this particular group is arbitrary because it has been established on the basis of all that happened during the past fifty-five million years. The horse family would not be a family if the first horse had been unsuccessful and had not given rise to a wide range of descendants. The origin of horses is therefore the origin of that first horse

Ceresiosaurus calcagnii, *middle Triassic of Monte San Giorgio, canton Ticino (Switzerland)*. Ceresiosaurus, *a reptile over two metres long, is one of the best known representatives of the nothosaurs — extinct reptiles related to the plesiosaurs and quite common during the Triassic. The large skeleton is surrounded by specimens of the smaller nothosaur,* Pachypleurosaurus edwardsi which, being a species, developed according to the rules of microevolution. On the basis of this example one can state that the large systematic categories have not necessarily appeared suddenly, in an evolutionary leap, and do not possess an ancestral prototype which might sum up all the characteristics of a given category.

Let us take another example. *Archaeopteryx*, the oldest known bird retains, as we shall see, some reptilian characteristics and has been considered to be half reptile and half bird. It is now classified as a bird because of its feathers and because the group of birds is now so widespread as to merit the level of a Class — the Class Aves — like the reptiles and mammals. Had birds not evolved to such an extent and had no other bird derived from *Archaeopteryx*, the latter would not have been classed separately and would have been considered a reptile, albeit a strange one, covered in feathers. Once again, therefore, the origin of a group is the same thing as the origin of the first representative and, consequently, of a species.

As was the case with some invertebrates, such as ammonites, it has been possible to throw some light on the phenomena which governed the evolution of even the discontinuous series of vertebrates. We have discovered the processes which led to evolutionary variations, such as migration, mutation, isolation and natural selection. They have all appeared quite clearly in the evolutionary history of the horse, a classic of palaeontological evolutionary evidence. Fossilized remains of horses have been found in many localities rather than in a single and continuous sequence of layers covering all the fifty-five million years of the horse's history. The reconstruction of such a history has therefore necessitated several correlations between considerably distant deposits and has revealed the existence of migrations, their direction and extent. We have thus been able to establish that, from an original form no larger than a hare and living in forests, horses have undergone a series of transformations which led them to their present dimensions and life in the prairies. All the various changes in dimensions, teeth (which adapted to feeding on grass rather than leaves) and feet (from four toes to one) took place together with a series of migrations between North America and Europe which are well documented by the fossils of both continents.

The evolutionary discontinuity which characterizes some larger systematic categories has been argued to be only apparent because of discoveries, during the 1920s, of certain evolutionary mechanisms which are linked to the development of the embryo or, more generally, to individual growth. These mechanisms can cause such fictitious discontinuities.

Ever since Haeckel's research, scientists have been aware of the parallelism between phylogenesis — the development of an evolutionary line — and ontogenesis — individual growth. It was thought that ontogenesis summarized phylogenesis in the sense that the growth of the individual would progress through embryonic stages along the lines of the evolutionary stages which had affected the phyletic line to which that particular individual belonged. Briefly, this meant that the evolution of a phyletic line would actually happen through the addition of new characteristics at the end of individual growth and that evolution itself was nothing but a sequence of adult individuals.

Bothriolepis canadensis, *Canadian Upper Devonian. One of the earlier types of jawed vertebrates, an "armoured" fish belonging to the extinct group of the Placodermi. Below: a reconstruction of the fish*

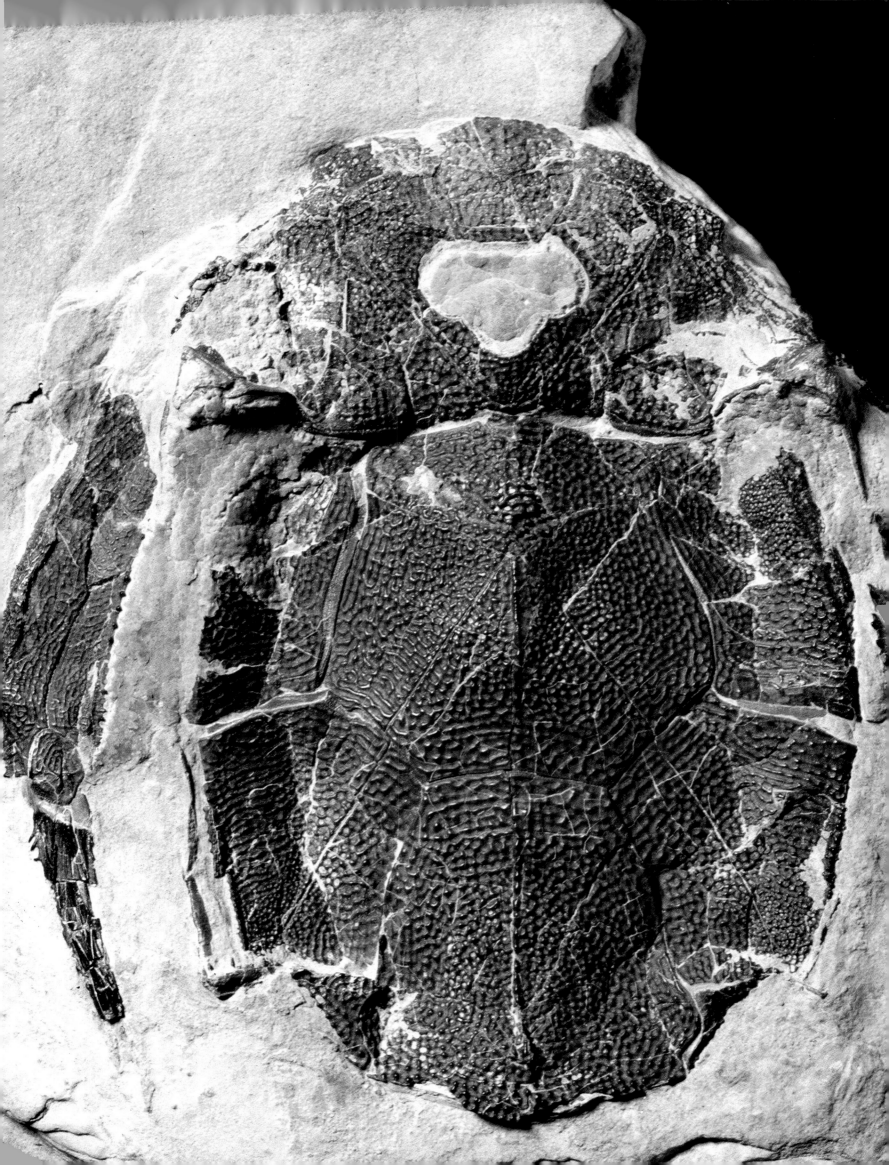

Important research carried out during this century has in fact shown that the relationship between individual growth and phyletic evolution is not so simple and that only rarely is evolution affected by the addition of new characteristics at the end of ontogenesis. New characteristics can be added at any time during individual growth, at the beginning, the end or half way through, thus producing different end-products according to a process which is today called heterochronism. When seen in this light, phylogenesis is no longer a successive sequence of adult stages, but a sequence of ontogenies with all the modifications which can take place during the ontogenies themselves.

The discovery of these mechanisms led to the demise of another basic biological concept, the one known as "the harmonious development of a type", according to which each organism is an integral system altered only by the perfect synchronization of all its parts. At the end of the nineteenth century, the palaeontologist Louis Dollo had become aware of the inconsistencies inherent in the harmonic development theory, and had surmised that the various organs of an individual could have independent phyletic histories: that they could, in other words, evolve independently from one another, with different speeds and in different ways. This principle is now known as "dissociability of the organs" and leads to the theory that an individual is only a mosaic of characteristics each of which can evolve according to its own rules. Dissociability and heterochronism — the phyletic individuality of organs and the addition of a characteristic in the course of ontogenesis — can produce interesting results with regards to evolution, and can cause those evolutionary leaps which had long been denied by evolutionists and had long been used by anti-evolutionists in their criticisms of the theory of evolution of the organic world.

It is worth looking at one of these processes, the one known as neoteny. Neoteny is the process in which an organism attains sexual maturity (and therefore reproductive ability) before having attained an adult body-form. In the course of the evolution of a phylum, this process causes entire late stages of development to be dropped, so that evolution can start anew on completely different premises.

Neoteny is shown most clearly in some amphibians which reach sexual maturity and become capable of reproduction while still living in the water and retaining gills and other larval characteristics. A neotenic process can cause an apparent evolutionary leap; such leaps have been used to explain the origins of man, of vertebrates and of flightless birds. The origin of vertebrates is an exemplary case. On the basis of certain shared characteristics which have been observed in the larval stages, it is thought that these highly evolved animals are closely related to echinoderms, such as starfish and sea-urchins which, when adult, show no affinity whatever with even the earliest vertebrates. However, the initial embryonic stages of vertebrates show distinct similarity to the larvae of echinoderms; it is on this basis that some authors have proposed that vertebrates might have originated as echinoderm-type larvae attaining sexual maturity through a process of neoteny. There would therefore be some evolutionary continuity between echinoderms and vertebrates, in so far as the two groups derived from similar larval stages, but there is no morphological continuity in the

Seymouria baylorensis, *Texas Permian*. Seymouria's *skeleton shows, in a kind of mosaic, both reptilian and amphibian characteristics. Although classified with the latter, it is considered to represent a link between the two vertebrate types and evidence of the evolutionary origin of reptiles from amphibians*

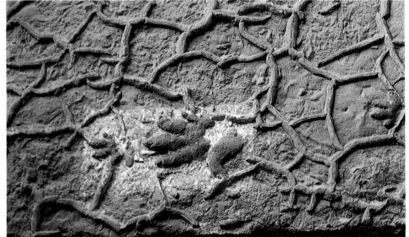

The marks left by a Chirotherium *on mud of the German Triassic. Fossilized footprints are very important to the study of our biological past as the oldest enable us to establish with some certainty in which period quadrupeds began to inhabit the dry land and how common they were. The two prints in the photograph have been attributed to a reptilian ancestor of the dinosaurs, the small ones being those of its fore legs and the larger ones of its hind legs. They are preserved in positive, like the ribs left in the mud by desiccation*

Rostra of belemnites, Holzmaden Jurassic (Germany). The belemnites are all that is left of the internal shell of a group of cephalopods now extinct but once analogous to modern squids

adult stages which are so different as to suggest a large gap in our knowledge of their history.

Link Fossils

Among the most interesting fossils to be found are those of organisms which possess the characteristics of two living groups which are now very different, and which appear to link major groups into evolutionary series. These "link fossils" can be thought of as transitional animals linking one life-style with another. Famous examples described below include an amphibious fish, a reptile-bird intermediate and the mammal-like reptiles. Link fossils are little known amongst fossil invertebrates, and the relationship between the major groups of invertebrates is still obscure. The great antiquity of the invertebrate phyla is the reason for the paucity of potential transitional fossils. At the beginning of the Cambrian, all the major groups of marine invertebrates were already in existence, diagnostic forms having been recognized in such fossil assemblages as the Burgess

Homotelus bromidensis, *trilobite of the Oklahoma Ordovician (United States of America). Trilobites are an extinct group of arthropods which were very common during the whole of the Palaeozoic*

Shale fauna from Canada. Transitional forms could thus only be found by reaching much further back in time, into the Archaeozoic which has so far yielded very few good fossils.

Within the vertebrates, however, the story is rather different, as they appear in the early Palaeozoic and are well documented in the fossil record. The evolution of vertebrates largely took place during geological periods which have yielded plenty of fossils, and it is therefore not surprising that some have been identified as link fossils which appear to be intermediates between the various classes: fish and amphibians, amphibians and reptiles, reptiles and both birds and mammals.

However, even in the case of the vertebrates, such link fossils are rare and there are two possible explanations for this. Firstly, the extreme speed of evolution when a group moves from one adaptive zone to another. Secondly, the possibility that innovations may have initially occurred in geographically restricted areas, so that the transition may have been very local in space and time. The link fossils identified among the fossil vertebrates include the amphibious fish *Eusthenopteron*, the primitive amphibian *Ichthyostega*, the reptile-like amphibian *Seymouria*, the primitive bird *Archaeopteryx* and the mammal-like reptiles or synapsids.

The Transition from Fish to Amphibian

The earliest known terrestrial vertebrates are the amphibians which first appeared in the Upper Devonian 350 million years ago. The living amphibians are the most primitive terrestrial vertebrates and have not completely abandoned aquatic life, most of them laying eggs in water and having an aquatic larval stage at the beginning of their life. Amongst the living fishes, the closest relatives of the amphibians are the Sarcopterygii represented by the coelacanth *Latimeria* (a deep-sea fish found in the Indian Ocean) and the lungfish which occur in fresh water in South America, Africa and Australia and which, like amphibians, have lungs and an internal skeleton to the fin. Palaeontologists wishing to find the closest fossil relatives of the amphibians therefore looked among the Sarcopterygian fishes which lived during the Devonian at about the time when the amphibians first appeared. The fossil fish which seems to be closest to the amphibians is an Upper Devonian form called *Eusthenopteron*, and the most primitive amphibian, also from the Upper Devonian, is a form called *Ichthyostega*.

Eusthenopteron was a carnivorous freshwater fish about 50 cm long, with a slightly elongated but essentially fish-shaped body. However, the fossils reveal that the arrangement of the bones in the skull, the structure of the backbone and the internal "arm" and "leg" skeletons within the paired fins all show great similarity to primitive amphibians. *Ichthyostega* was at the other end of the transition, being an undoubted amphibian with a few retained fish-like features, such as fin-rays in the tail. As well as the construction of the skull and backbone, the two animals both had teeth with a characteristic labyrinthine structure, i.e. with complex infoldings of the outer layers of enamel and dentine to give internal strengthening struts which in section look like a labyrinth. The major group of early amphibians is sometimes called the Labyrinthodontia because of this tooth construction.

These two link fossils shed some light on a fundamental part of the history of vertebrates, indicating that the fish-amphibian transition took place prior to the Upper Devonian in fresh water, initially giving rise to the labyrinthodonts.

The Transition from Amphibian to Reptile

The oldest known reptiles have been found in Carboniferous rocks about 310 million years old. Reptiles are more specialized to a terrestrial existence than amphibians and can live independently of water, mainly because of the innovation of the amniotic egg, which does not need to be laid in water since it has (inside a protective shell) a supply of food in the form of a yolk-sac. Fossilized eggs have been found, although the oldest are in Triassic rocks which are much more recent than those containing the first reptiles. Apart from the amniotic egg, other reptile innovations which can be seen in the skeleton, and hence in fossils, include the reduction of some of the bones at the back of the skull, the absence of a semicircular notch supporting the large ear-drum typical of amphibians, and the fusion

Dactylioceras athleticum, an ammonite of the Lower Jurassic of Whitby, Great Britain. Like the trilobites, the ammonites are also extinct; the only organism bearing some sort of similarity to them is the cephalopod Nautilus

On the following page: a much enlarged specimen of Smerdis minutus *of the Eocene of Aix-en-Provence (France). It still retains many details of the skeleton and the coloured mark of the soft parts*

of the sacral vertebrae to the pelvis. Most amphibian or reptile fossils can be readily identified in one way or the other but there are several Carboniferous and Permian families which are difficult to assign. They are believed to be amphibians of the Order Anthracosauria, as one form is known to have had aquatic larvae, but they are similar to early reptiles and would probably have been identified as reptile ancestors had they not been slightly later in occurrence than the earliest true reptiles.

One of the best known anthracosaurs is *Seymouria*, first described as a possible reptile but now known to have been a terrestrial amphibian which lived when reptiles were already widespread in the Lower Permian. It was about 80 cm long and had a typical early amphibian skull with a large ear notch and labyrinthine teeth combined with a skeleton more like that of an early reptile in the strongly constructed backbone and the fused sacral vertebrae. Although too late to be genuine transitional forms, *Seymouria* and the other anthracosaurs appear to be a series of relicts of the amphibian-reptile transition which must have taken place in the Lower Carboniferous.

The Transition from Reptile to Bird

During the Jurassic, reptiles reached the peak of their diversity, living in practically all environments. There were terrestrial reptiles, freshwater reptiles, marine reptiles and even some flying reptiles, the pterosaurs. These strange flying creatures had conquered the skies in the Triassic thanks to the modification of their fore limbs as wings, each wing being made up of a membrane of skin, supported along its leading edge by a single greatly elongated finger of the hand. Pterosaurs had several analogies with birds, they had pneumatized bones (hollow bones containing air-sacs which are extensions of the lungs) and a skull lightened by several large openings. Not surprisingly therefore, some early palaeontologists believed them to have been the predecessors of modern birds.

All this was changed by the sensational discovery in the mid-nineteenth century of a remarkable fossil. The lagoon limestones of the Upper Jurassic at Solnhofen in Bavaria (Germany), yielded an intact fossil of a strange creature which was named *Archaeopteryx lithographica*. Four more specimens of *Archaeopteryx* have come to light subsequently. From its first discovery, it was immediately apparent that it had both reptile and bird characteristics. It is now considered to be the most primitive known bird retaining many characters of its reptilian ancestry.

Archaeopteryx was a pigeon-sized bird and possessed feathers, wings and the characteristic bone of the shoulder girdle, the furcula or "wishbone". The feathers are preserved only because of the unusually fine sediments which make up the Solnhofen limestone. The reptilian characteristics retained in *Archaeopteryx* include pointed teeth in sockets, a skull constructed like those of some small dinosaurs and thecodont reptiles, the absence of pneumatic cavities in the bones, a long bony tail with feathers implanted all along its length, and some of the fingers still unmodified for flight and bearing large grasping claws. The structure of *Archaeopteryx* leads to the inevitable conclusion that it is the most primitive known bird and that its closest

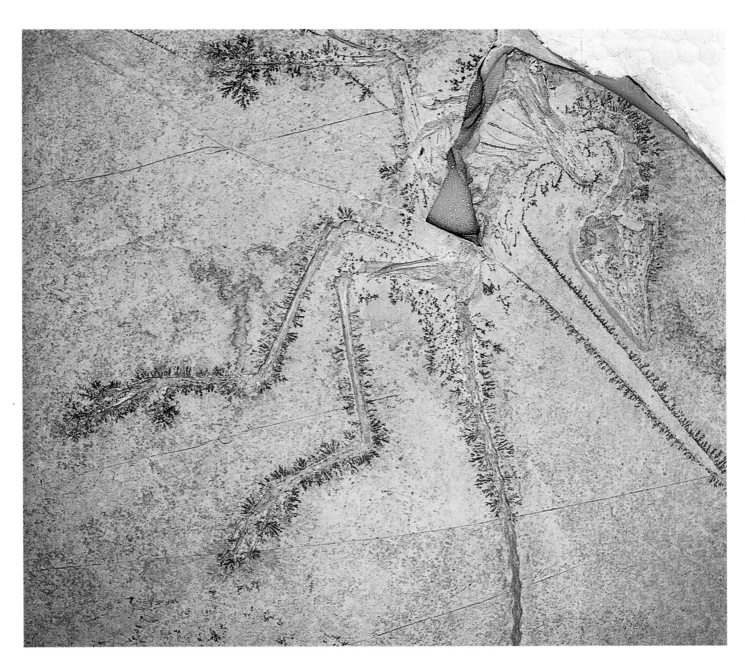

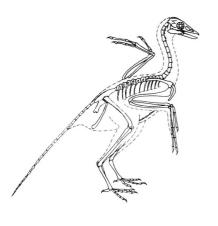

Above: *one of the five known specimens of* Archaeopteryx lithographica *of the Eichstatt Upper Jurassic.* Right: *an osteological atlas of the specimen.* Left: *a reconstruction of the skeleton. The Archaeopteryx is the oldest bird we know, having feathers as well as many reptilian characteristics. It demonstrates the evolutionary origin of birds from reptiles, more precisely from a group of saurischian dinosaurs*

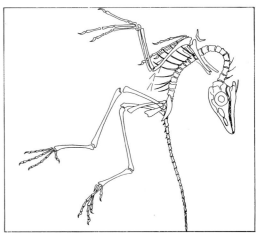

reptilian relatives are either some of the early small carnivorous dinosaurs or their thecodont precursors.

The feathered bird wing appears to have been a much more versatile structure than the pterosaur wing, capable of functioning in a wide range of shapes and sizes. Either for this, or for some as yet unknown reason, the birds appear to have replaced the small and medium-sized pterosaurs during the Cretaceous. Only the largest pterosaurs survived to the end of the Cretaceous when they became extinct suddenly, along with many other life-forms.

The Transition from Reptile to Mammal

The previous examples have all involved forms presenting transitional characteristics of the two groups which they link. The transition from primitive reptile to mammal is, however, decidedly more complex, because of the abundance of material. The reptilian Subclass Synapsida, found in rocks from Upper Carboniferous to Middle Jurassic age, comprises many families which show a wide spectrum of skeletal structure from primitive reptile to near-mammal.

The synapsids which bear the closest resemblance to early mammals and

A colony of Archimedes *bryozoans with its characteristic helicoidal formation, found in marine sediments of the Lower Carboniferous of Connecticut (United States of America)*

The crinoid Hypselocrinus *(white in the photograph) together with bryozoan colonies in the rocks of the Lower Carboniferous of Connecticut (United States of America)*

which are believed to be their closest relatives are the Cynodontia, known from the Triassic of most continents but particularly from Africa, South America and Russia. They ranged from 50 cm to 2 m in length and had a mammal-like dentition and jaw construction, and a mammal-like posture and limb construction. However, they possessed tiny brains (as can be seen from the shape of the skull) and the backbone bore ribs almost back to the pelvis indicating that there was no diaphragm to aid efficient breathing as in mammls. Because there is a continuum of structure from the cynodonts through the early mammals, defining a reptile-mammal boundary is an exercise which is inherently arbitrary and (because of the small size and fragmentary nature of many early mammals) very difficult for all but the best material. The early mammals appear to have been a lineage of "dwarf" cynodonts and the structural changes associated with this reduction to an adult size of about 20 cm are those which are most widely used to define early fossil mammals. One widely-used character has been the change in function of the reptilian jaw-hinge bones, the quadrate and articular, into tiny sound-transmitting bones, the incus and malleus in the middle ear of mammals. This is a single skeletal manifestation of extensive changes which took place in the jaws and ears of these forms and there is a risk in using just one character — it may have evolved more than once. In the past it has been suggested that two groups of mammal-like reptiles may have independently evolved to a mammalian structure, but consideration of many characteristics makes this appear unlikely and it is no longer widely believed. The small size and spiky teeth of the earliest true mammals suggests that they were small, possibly nocturnal, shrew-like forms. If so, this would correlate with the mammalian characteristics of a furry covering and a high body temperature and metabolic rate. A small animal hunting for food in a cool night, requires insulation and the ability to maintain its metabolic rate and temperature independent of the solar heat required by reptiles. This transition of life-style must have occurred either in the Middle or the Upper Triassic.

The Appearance of New Structures

Fossil remains have been found which, while not representing real links between different groups of animals, nevertheless possess anatomical characteristics which suggest how a particular structure or new organ may have first appeared. A good example of this is supplied by the genus *Acanthodes*, a primitive fish (sometimes called a "spiny-shark" although it is not a true shark) which shows how the lower jaw may have originated. The jaw is actually the product of the modification of certain skeletal elements next to the skull, which had no part in feeding, namely the branchial arches. The branchial arches are cartilaginous supports, in the shape of a V lying on its side. They are placed in rows between the gill openings and are used to strengthen this part of the body. *Acanthodes* shows how the first branchial arch developed and changed, and finally formed the upper jaw, or palate, while the second arch formed the lower jaw or mandible. Without this fossil discovery, such a transformation would probably be still unknown.

The Geological Eras

ERA	PERIOD		DATE WHEN STARTED (IN MILLION YEARS)
Quaternary	Holocene		
	Pleistocene		2
Cenozoic or Tertiary	Neogene	Pliocene	
		Miocene	22
	Paleogene	Oligocene	
		Eocene	
		Palaeocene	65
Mesozoic or Secondary	Cretaceous		140
	Jurassic		195
	Triassic		230
Palaeozoic or Primary	Permian		280
	Carboniferous		345
	Devonian		395
	Silurian		435
	Ordovician		500
	Cambrian		570
Archaeozoic	Algonkian		2,600
	Archaean		4,500

The division of the history of the Earth into eras, periods and lesser sub-divisions is a necessary aid to the scientific knowledge of the past. It has enabled scientists to reconstruct, as far as possible, the past of our planet by means of correlations of faunas and floras, as well as of geological and geographical events. This device, which today seems so obvious, was invented by Giovanni Arduino (1713–95), a professor of mineralogy at Padua in Italy. He was the first to divide the sequences of rock layers of the Pre-Alps into three groups which he called "Primary", "Secondary" and "Tertiary", respectively from the oldest to the most recent.

The Palaeozoic Era

PERIOD OR SYSTEM	EPOCH OR SERIES	AGE	DATE WHEN STARTED (IN MILLION YEARS)
Permian after Perm, USSR	Zechstein	Thuringian	250
	Rothliegende	Saxonian	
		Autunian	280
Carboniferous after the coal deposits	Pennsylvanian	Stephanian	310
		Westphalian	
	Mississippian		345
Devonian after Devon, British county	Upper Devonian	Famennian	360
		Frasnian	
	Middle Devonian	Givetian	370
		Eifelian	
	Lower Devonian	Emsian	395
		Siegenian	
		Gedinnian	
Silurian after the Silures, inhabitants of Wales in Roman times	Upper Silurian	Ludlovian	420
		Wenlockian	
	Lower Silurian	Llandoverian	435
Ordovician after the Ordovices, inhabitants of Wales in Roman times	Upper Ordovician	Ashgillian	450
		Caradocian	
	Lower Ordovician	Llandeilian	500
		Llanvirnian	
		Skiddavian	
		Tremadocian	
Cambrian after the Roman name for Wales	Upper Cambrian	Potsdamian	520
	Lower Cambrian	Acadian	570
		Georgian	

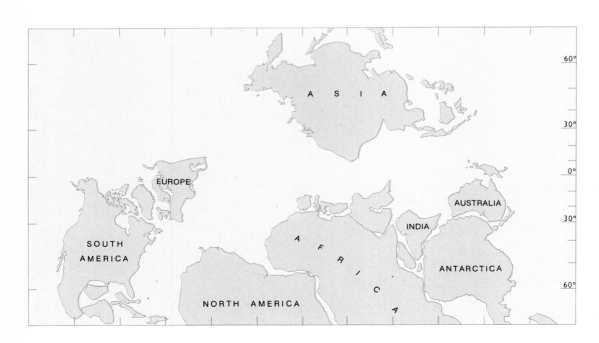

The position of the continents during the Carboniferous, some 340 million years ago. Europe was welded to North America, the southern continents formed a single mass separated from the others, and Asia was isolated. The merging of Asia and Europe later led to the formation of the Urals

The Mesozoic Era

PERIOD OR SYSTEM	EPOCH OR SERIES	AGE	DATE WHEN STARTED (IN MILLION YEARS)
Cretaceous after *Creta*, the Latin word for chalk, characteristic of the era and appearing in France, Germany, Great Britain and Belgium	Upper Cretaceous	Maastrichtian	100
		Senonian	
		Turonian	
		Cenomanian	
	Lower Cretaceous	Albian	140
		Aptian	
		Barremian	
		Neocomian	
Jurassic after the Jura mountains, the rocks of which are widespread	Malm	Tithonian	160
		Kimmeridgian	
		Oxfordian	
	Dogger	Callovian	176
		Bathonian	
		Bajocian	
	Lias	Aalenian	195
		Toarcian	
		Pliensbachian	
		Sinemurian	
		Hettangian	
Triassic represented by three groups of different soils in central and western Europe	Keuper	Rhaetian	215
		Norian	
		Carnian	
	Muschelkalk	Ladinian	225
		Anisian	
	Buntsandstein	Scythian	230

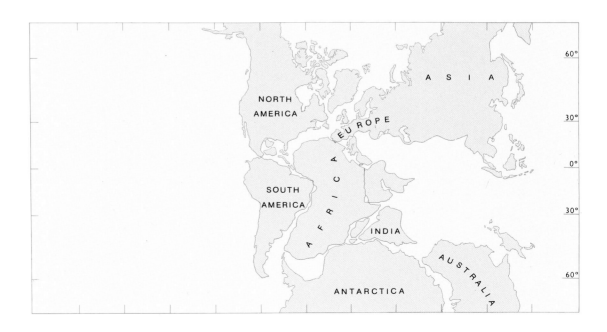

The continental picture during the Triassic, about 220 million years ago. The continents formed a single mass, broken on the eastern side by a deep and vast oceanic gulf, the Tethys

The Cenozoic Era

PERIOD OR SYSTEM	EPOCH OR SERIES	AGE	DATE WHEN STARTED (IN MILLION YEARS)
Neogene after the Greek for "new birth"	Pliocene	Piacenzian	5
		Zanclean	
	Miocene	Messinian	22
		Tortonian	
		Langhian	
		Burdigalian	
		Aquitanian	
Palaeogene after the Greek for "ancient birth"	Oligocene	Chattian	40
		Rupelian	
	Eocene	Priabonian	55
		Lutetian	
		Ypresian	
	Palaeocene	Thanetian	65
		Montian	
		Danian	

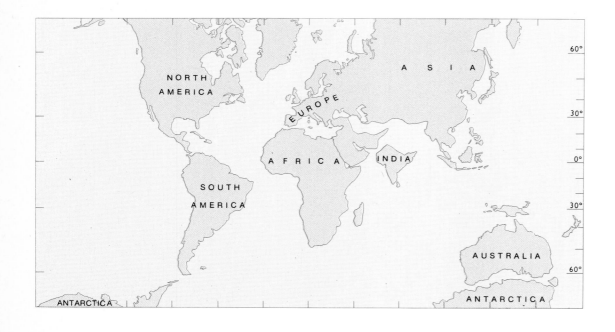

The continents during the Eocene, some 50 million years ago. The modern pattern is almost there, except that Antarctica is still connected with Australia, and India has not yet joined Asia; the fusion of the latter two continents will lead to the formation of the Himalayas

The Quaternary Era

GEOLOGICAL DIVISIONS		CULTURAL SEQUENCES		HUMAN TYPES	BEGAN (YEARS AGO)
HOLOCENE	POST-GLACIAL	Acquisition of productive economy, appearance of settlements, agriculture, breeding animals, weaving, ceramics, stone wares	NEOLITHIC	Homo sapiens sapiens	6,000
		MAGLEMOSIAN TARDENOISIAN AZILIAN	MESO-LITHIC	(Brachycephalous types are widespread)	9,000
UPPER PLEISTOCENE	WURM	MAGDALENIAN SOLUTREAN GRAVETTIAN AURIGNACIAN CHATELPERRONIAN	UPPER PAL.	Neoanthropinae Cromagnon Chancelade Grimaldi	50,000
		TAYACIAN / LEVALLOIS TECHNIQUE / MOUSTERIAN / MICOQUIAN	MIDDLE PAL.	Palaeoanthropinae Neanderthal Homo sapiens neanderthalensis	100,000
	RISS		LOWER PALAEOLITHIC	early Homo sapiens	300,000
MIDDLE PLEISTOCENE	MINDEL	CLACTONIAN ACHEULIAN			
	GUNZ	ABBEVILLIAN		Homo erectus	1 million
LOWER PLEISTOCENE	DONAU	Oldowan pebble culture, choppers and chopping tools		Homo habilis	1,850,000
				Australopithecus africanus robustus and afarensis	3 million

The Palaeozoic Era

The Palaeozoic Era

The Palaeozoic era began about 570 million years ago and lasted some 340 million years. It is the first era in the history of the Earth to yield a rich enough documentation for us to reconstruct the main geological and biological events. The Palaeozoic followed the birth of the Earth itself by many million years, roughly four billion years. Very little is known about this huge interval — called the Archaeozoic; few fossils have survived and geological evidence is limited. The Palaeozoic, on the other hand, has handed down to us large numbers of fossils, and its geological strata allow us to reconstruct the geography of the planet throughout its various periods.

The Earth's geography underwent many changes: the continental masses, clearly separate at the beginning of the era, shifted considerably until, during the Permian period, they joined together to form a single continental mass now called Pangaea. Living organisms rapidly evolved on these slow-moving continents and in the seas between them. The main stages of this evolutionary process are recorded in Palaeozoic rocks: they are the development, during the Cambrian, of the first aquatic vertebrates; the conquest of dry land during the Silurian period on the part of plants and invertebrates; the first appearance of amphibians, during the Devonian, and the conquest of dry land by the vertebrates; and, during the Carboniferous, the appearance of the very first reptiles.

The star-like shapes on this grey rock are called Dactyloidites. *They are Cambrian fossils, found in the United States, which are thought to be algae or the prints of the mantle of primitive medusae. The latter were very common in the Cambrian: notwithstanding their gelatinous body (ninety per cent water), they have often left very clear fossilized traces*

The Palaeozoic, also called the Primary era because it was once believed that no older rocks existed, first started, according to geologists, 570 million years ago and came to an end about 230 million years ago. It therefore lasted some 340 million years — an extremely long time, during which took place both the geological and the geographical changes which altered the face of the Earth several times, and the consequent rapid evolution of living organisms. It was indeed during this era that the latter achieved some of their most important evolutionary advances. Modern research methods allow geologists to reconstruct several of these events fairly easily, while palaeontologists, aided by constant new discoveries of fossils, have been able to draw a fairly accurate map of evolutionary conquests and changes in both the animal and the vegetable worlds. As a result, our knowledge of the era can now be said to have achieved a degree of accuracy — something which was unimaginable only a few years ago.

The beginning of the Palaeozoic represents the furthest limit of our knowledge of the past. Whatever preceded it, both biologically and geologically, is shrouded in a mystery rendered obscure by the shortage of precise data. With the beginning of the era, on the other hand, the information is so plentiful that any reconstruction of the past is, if not easy,

Specimen and fragments of Cryptoblastus melo, *of the North American Carboniferous. It is the outer shell of a blastoid, an echinoderm similar to modern crinoids, or sea lilies, but belonging to an extinct group*

The mark left by a shell of the genus Orthis *on a Silurian rock found in England. Brachiopods were very common during the whole of the Palaeozoic. Their tough shell has enabled them to fossilize almost perfectly*

The coral Halysites catenularia, *shown here, is so called after the chain-like disposition of the cups. It was a colonial coelenterate, one of the tabulate corals which, together with many other types, contributed to the formation, in the Silurian, of the first coral reefs. This specimen was found in England*

certainly less difficult or dubious. The various geological and biological events which took place during the Palaeozoic allow us to divide it into shorter periods, thus greatly facilitating scientific research. Each period is determined by more uniform and typical characteristics of fauna and flora. On the basis of such divisions, links are possible between contemporary deposits found at great distances. The result is a reconstruction of the Earth as it appeared in the various periods of its history, and, consequently, of its evolution.

The Palaeozoic is divided into six periods of unequal length, named either after the localities in which the various deposits are better represented or were first found, or after the peoples who lived there in the past, or after certain specific characteristics. The Cambrian, for instance, the first period of the era, is named after Cambria, the name the Romans gave to Wales; the Ordovician after the Ordovices, ancient inhabitants of Wales; the Silurian after the Silures, who also inhabited Wales; the Devonian after the British county of Devon; the Carboniferous after the vast coal deposits which formed during the period all over the world; and the Permian after the district of Perm, in the Urals.

Geologists have calculated that the maximum thickness of all the sediments formed during the 340 million years of the Palaeozoic is about 30 km, as in the case of marine and continental rocks such as desert sand or aquatic silts which testify to the geographical and biotopical evolution of the Earth.

Geographical Evolution

At the beginning of the Palaeozoic, during the Cambrian, the continents we are familiar with today did not exist. In their place there were four continental masses separated by deep seas and corresponding to continental Europe, North America, continental Asia and a mass formed by the union of what today are South America, Africa, Australia, Antarctica, India and Madagascar.

The position of these continental masses on the surface of the planet was also different. It appears that Africa, South America and central Asia have since made an about-turn of about 180° on their axis. The continents were separated by deep sea basins within which huge quantities of sediments were settling. These sediments would give rise, as a result of a series of orogenic movements, to massive mountain ranges. One of these ocean basins was to be particularly important to the evolution of the Earth's

Above: *the perfectly preserved head of this trilobite of the species* Dalmanites caudatus *shows the kidney-shaped eyes. It has been demonstrated that the eyes of these arthropods were almost perfect optical instruments.*

Below: *a small complete specimen of the species* Calymene blumenbachii, *a trilobite characteristic of the Silurian. The specimen, found in English deposits, is in the initial stage of coiling itself up, a defence mechanism typical of trilobites*

Facing page: *a large and perfect specimen of* Phacops rana, *a trilobite of the North American Devonian. The specimen clearly shows the partitioning of the body both transversely and longitudinally, a partitioning which caused them to be called trilobites*

Wadi Rum

Graptolites, once regarded as plant remains or fragments of cephalopods and coelenterates, are actually fossilized stomochordates. They are animals which occupy an intermediate position between the invertebrates and the chordates, the latter being the group to which vertebrates belong. Although it is very difficult to observe, with the naked eye, any affinity between graptolites and chordates, the presence of a stomochord analogous to the notochord of chordates indicates a certain kinship

These rocks surfacing in the southern desert of Jordan, not far south of Wadi Rum, date from the Silurian. Their layers have yielded several graptolites, three of which are illustrated. These animals lived mainly in the Silurian; they formed large pelagic colonies which were carried along by currents, thus achieving considerable distribution.

geography: the one which separated continental Europe from North America and which, later on, rose to weld the two continents together.

The Cambrian continental masses had not totally emerged; parts were covered by shallow seas which deposited sediments rich in organic remains. Elsewhere (in the strictly continental areas), lacustrine, marshy and aeolian sediments formed which contain no fossils, thus indicating that life was then confined to the seas. The climate was the same for all continents, neither particularly warm nor particularly cold.

This geographical situation lasted for 130 million years, until the beginning of the Devonian. Around this time, available data indicate substantial geographical changes. Towards the end of the Silurian, a widespread orogenic movement, the Caledonian orogenesis, led to the formation of new mountain ranges and the union of the North American with the European continents. During the Devonian, therefore, modern continents were gathered in three masses, separated (like the previous ones) by deep sea basins rich in sediments.

These continental masses, well known to geologists, are today called Euramerica (formed by North America and Europe), Angara (formed by continental Asia), and Gondwanaland (formed by all the southern lands and India).

Continental sediments so far discovered seem to indicate that the climate of the Devonian was periodically arid. The red sandstone formations which surface in Europe and North America were probably deposited in desert areas, swept by the wind and dotted with swamps of various dimensions. Plant fossils of a continental type appear in great numbers in these swamps and indicate the first colonization of dry lands. During the later Carboniferous period, the Earth's geography changed again. The three continental masses moved closer to one another, initiating a process which led in the Permian, to the formation of a single mass of dry land; and the Tethys Ocean — the deep sea which for millions of years divided the southern from the northern lands — also began to appear.

The Carboniferous was humid and warm in equatorial regions; large forests covered the dry lands and the accumulation of their remains formed the coal deposits which can now be found all over the world, from North America to Europe and China.

These equatorial regions, so rich in vegetation, were ideally suited to a flourishing life, as is clearly shown by the discoveries, in continental Carboniferous rocks, of large numbers of land animals: amphibians, reptiles and a wide range of insects.

The last period of the Palaeozoic witnessed another geographical change. During the Permian, the movement of the three continental masses towards one another came to its last stage and led to the formation of a single continent consisting of all the known dry lands; a supercontinent, called Pangaea by geologists. This continent was irregularly shaped: on the eastern side it had a huge marine gulf — the immensely deep Tethys — into which flowed and accumulated the sediments being eroded from newly-formed mountain ranges. This sedimentation process is an extremely important event which went on for many periods. Once lifted and

A complete specimen of Drepanaspis gemuendensis *of the Lower Devonian, Germany. This was another "armoured" fish belonging to the group of the agnathans, the oldest vertebrates we know, having neither jaws nor paired fins. This fossil shows quite clearly the bony plates which covered the fish, and the eyes and the tail covered in small platelets.*

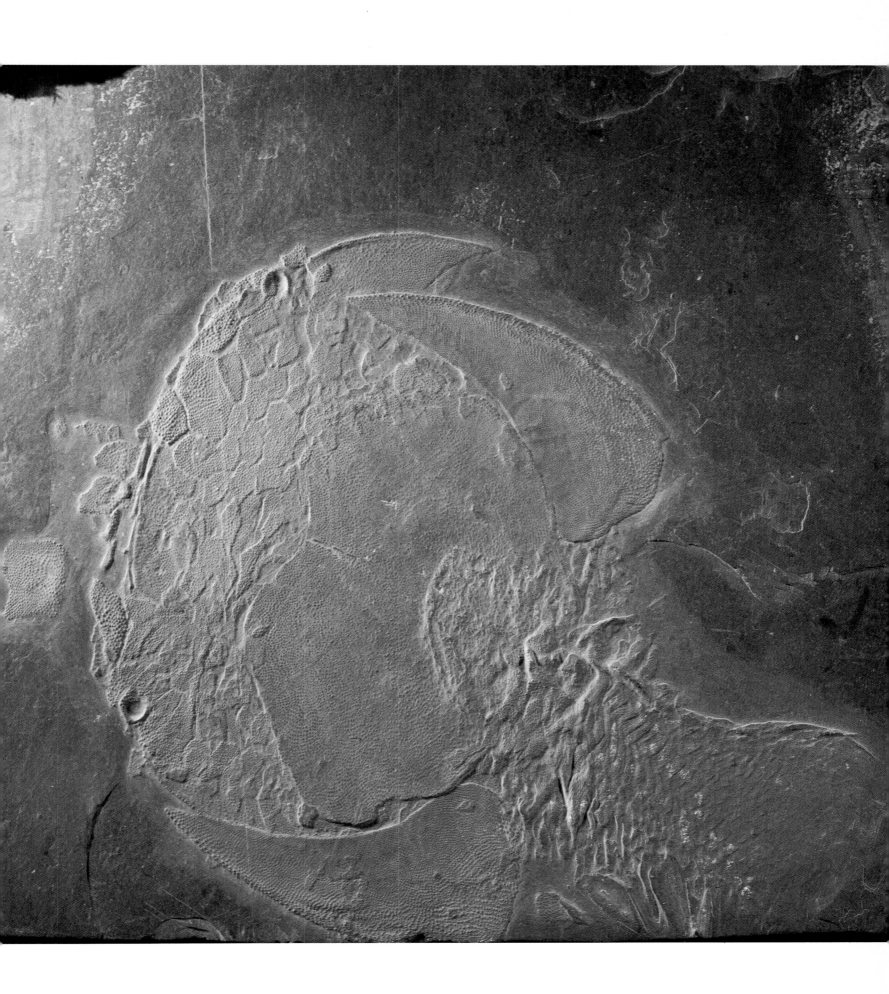

Above: *crossopterygian fishes lived in this sort of habitat, with shallow and marshy waters. The first terrestrial vertebrates, the amphibians, are believed to have originated from these fishes during the Devonian*

Below: *this specimen of* Pterichthyodes, *from the Scottish Devonian, belonged to a group of ancient vertebrates which had evolved further than the agnathans. It is a placoderm, a fish with jaws and paired fins which enabled it to swim more efficiently*

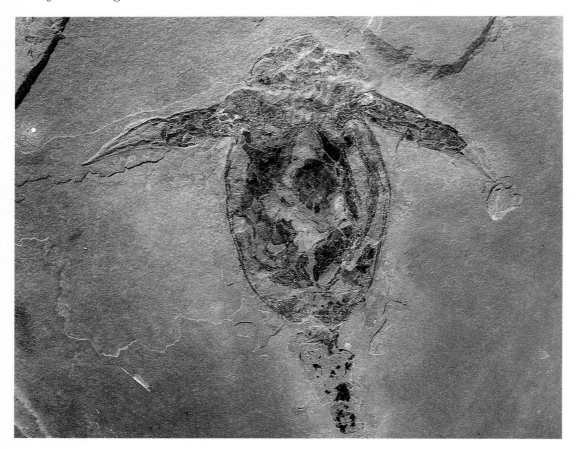

Right: *this specimen of* Gyroptychius *represents a very interesting group of fishes, the crossopterygians (Crossopterygii) from which the amphibians originated during the Devonian. The crossopterygians had vertebral and limb structures which were relatively similar to those of the most primitive amphibians. (Middle Devonian, Scotland)*

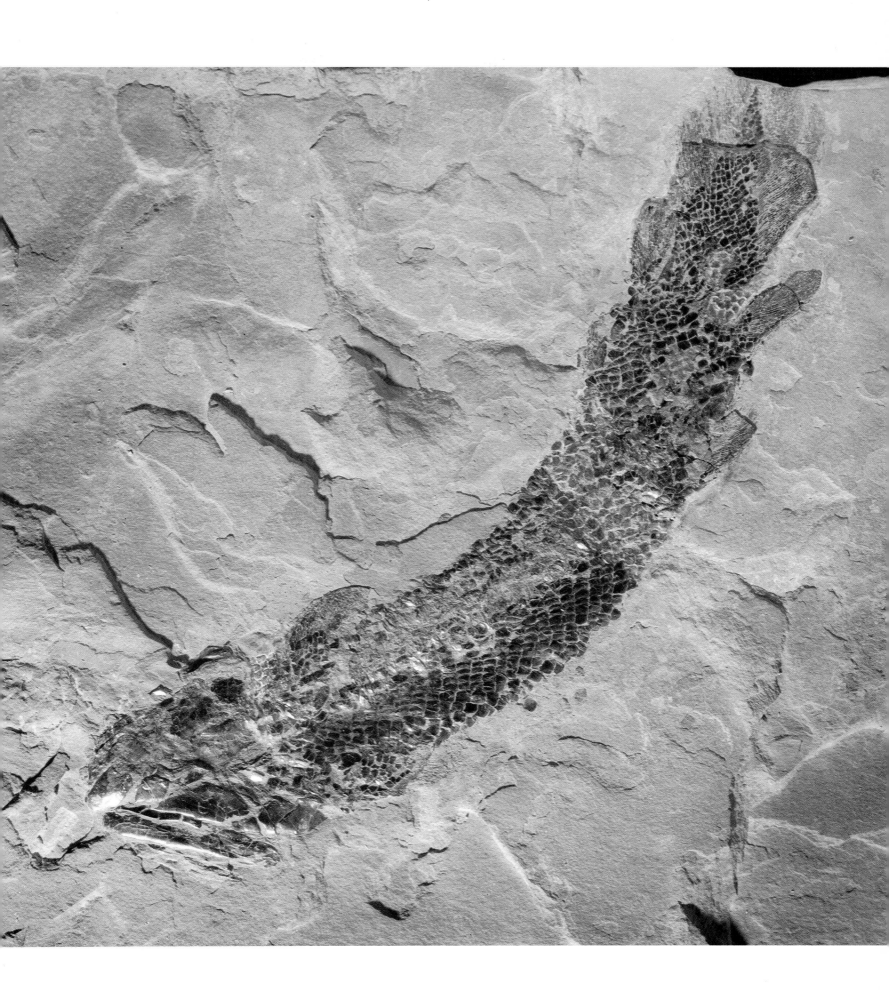

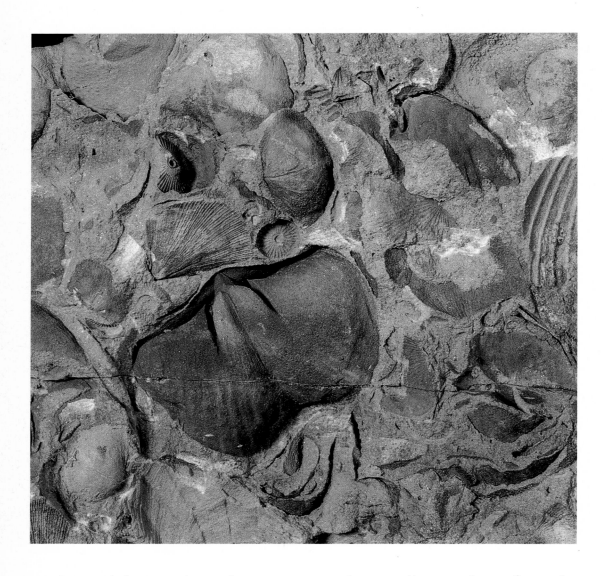

compressed by continental movements, these sediments later formed, many million years after the Permian, the massive mountain ranges which were born during the alpine orogenesis.

Pangaea was subject to substantial climatic variations. In the southern lands signs have been found of an extensive glaciation, followed by a warm and humid climate. In the north, the climate was dry, occasionally desert, as shown by the vast expanses of sandstone, of aeolian origin, and the saltpetre formations common in Europe and North America. Such a difference in climate was bound to affect both flora and fauna, which developed separately and produced a southern flora and fauna with their own characteristics, and a northern flora and fauna which, although more limited, were nonetheless just as interesting from the evolutionary perspective.

As we have seen, the fossils found in such quantity in marine and continental sediments of the Palaeozoic have allowed us to reconstruct fairly comprehensively the stages of organic evolution which took place during this period. This is very important as the organisms of the Palaeozoic made considerable steps forward: they conquered biotopes different from the marine one, to which they were confined at the

Casts of the interiors of brachiopods in a Belgian Carboniferous rock. The largest of the specimens is a Productus, *a zone-fossil of the Carboniferous. In this period, while the continents were covered in luxuriant forests, the oceans were full of the sort of life we find today in tropical waters*

The footprint of a seymouriamorph amphibian which lived in Europe in the Lower Permian, about 280 million years ago. These fossilized amphibians show certain characteristics which are transitional to the primitive reptiles and are therefore regarded as very close to the latter's ancestors

beginning of the era, colonized the dry land and spread over all the continents in more highly evolved forms.

The Evolution of Life

The beginning of the Palaeozoic, the Cambrian period, is represented only by marine fossils, indicating that life was largely, if not completely, restricted to the sea. However, a vast range of marine organisms was present. As well as containing seaweeds and reef-forming algae (stromatolites), the Cambrian seas were inhabited by all the major groups of invertebrates known today, and by others now extinct. There were siliceous sponges, jellyfish, annelid worms, brachiopods of various shapes and sizes, primitive echinoderms, early lamellibranch, gastropod and cephalopod molluscs and an immense variety of arthropods, in particular the trilobites.

Extinct groups include the archaeocyathans (halfway between sponges and coelenterates) and the graptolites (colonial organisms possibly related to the chordates). Possibly the most interesting Cambrian fauna was discovered at the turn of the century on Mount Wapta in British Columbia

(Canada), where the dark Burgess Shale, produced as the result of the collapse of a submarine mud-cliff, contains a beautifully preserved fauna of soft-bodied marine animals. Delicate worms are preserved in minute detail, having been buried intact in non-decomposing conditions. Trilobites and early crustaceans are preserved with all their legs, mouthparts and antennae intact and can thus be reconstructed in more detail than those from any other Cambrian locality. This one locality has probably revealed more about the diversity of life in the Cambrian than all other localities representing this period.

One group of organisms not represented in the Burgess Shale, but which also appeared in the late Cambrian, some 510 million years ago, is the Subclass Vertebrata. The first vertebrates have been found in Upper Cambrian and Lower Ordovician beds in the Rocky Mountains. The fossils consist only of isolated armour plates and scales of primitive jawless "fishes"; thus, little is known of the shape or lifestyle of these early forms. The remains are sufficiently scarce, and probably transported, that it cannot even be ascertained whether they are marine or freshwater in origin.

The Ordovician flora and fauna are not substantially different from those of the Cambrian, apart from the appearance of the planktonic graptolites. These colonial organisms lived suspended in sea-water, and are found in abundance in Ordovician and Silurian rocks.

The vertebrates became widespread during the late Ordovician and Silurian, both seas and fresh waters of the latter period being populated by vertebrates belonging to the Class Agnatha. They all lacked jaws and many lacked fins and must have wriggled through the water rather like tadpoles. Many were convered in thick armour plates which fossilize well and were undoubtedly a device to protect the early agnathans. In the same waters lived the largest of all known arthropods — the giant water-scorpions, predators up to 2 m long which must certainly have preyed on the small vertebrates.

From a biological viewpoint, the Silurian is most noteworthy as the period when living organisms conquered the dry land. The presence of fossils of a few genera of vascular plant shows that during this period, some 420 million years ago, the continents acquired their first covering of vegetation, almost immediately followed by the arrival of invertebrates. The earliest land animals are found in Silurian rocks: they are millipedes found in Welsh rocks, and a scorpion, *Palaeophonus nuncius*, found in the island of Gotland (Sweden).

The available information indicates that terrestrial plants became much more diverse during the Devonian with the appearance of the Psilophytales (primitive swamp plants) early in the period and the later appearance of horse-tails (Equisetales), club-mosses (Lycopodiales) and ferns (Pteridophyta). Such a rich vegetation led to a much more substantial colonization of the dry lands than had taken place in the Silurian. Small wingless insects and mites are known from the Lower Devonian of Scotland and centipedes and tarantulas have recently been discovered in the Middle Devonian of New York State. Towards the end of the Devonian, the earliest known land vertebrates appear, as witnessed both

During the Carboniferous the continents were covered with immense forests consisting of tree ferns, lycopods, horse-tails and gigantic pteridosperms. Coal deposits are all that is left of these forests and they often yield the fossilized remains of the original plants; see for instance the fragment of a lycopod of the genus Lepidodendron *in the inset*

Left: *a fern from the Upper Carboniferous of Germany,* Acitheca polymorpha. *The perfectly preserved branches show the tiniest veins*

Right: *another plant which lived in the forests of the Carboniferous,* Senftenbergia plumosa. *This specimen was found in the coal deposits of the Saar*

Below: *detail of the trunk of a* Sigillaria, *a lycopod. The largest of the lycopods could be as much as 30 metres tall. This specimen, found in the Carboniferous deposits of Belgium, is red because, during the fossilization process, it became impregnated with iron minerals.*

by footprints from Australia and by the remains of several types of primitive amphibian from Greenland, most notably *Ichthyostega*. *Ichthyostega* was about 1 m long with large limbs fully adapted to walking on land. It is however, significantly more primitive than any Carboniferous amphibian. Its large pointed teeth suggest that it was a predator on either fishes or very large arthropods.

While some vertebrates were moving onto the land, other forms were developing in the Devonian seas: at the beginning of the period the placoderms, the sharks and the bony fishes all appeared; all forms with jaws and paired fins. The armoured placoderms were exceptionally diverse during the Devonian. They were the most primitive of the jawed vertebrates and were replaced by more advanced fishes by the end of the period. In the same seas, the trilobites remained the most abundant arthropods and a group of molluscs appeared which was to spread considerably in successive periods: the ammonites.

The Devonian was succeeded by the Carboniferous which, in equatorial regions, was undoubtedly the most luxuriant period, the one richest in life, of the whole of the Palaeozoic. The Carboniferous world was characterized by a warm and humid climate on many continents favouring the growth of lush vegetation. Vast swamp-forests of horse-tails (*Calamites*) and giant

club-mosses up to 30 m tall, together with pteridosperms and tree-ferns, provided an ideal environment for rich tropical fauna. Some of the stagnant pools in these swamps provided suitable conditions for this rich fauna to be preserved. Spiders, scorpions, millipedes, woodlice and hundreds of species of winged insect lived in these forests and were preyed on by the many kinds of amphibian which lived both in the pools and on the land. At this time, the amphibians achieved a peak of diversity in shape and size, which they would never repeat, there being forms from a few centimetres to 4 m in length, crocodile-like amphibians, snake-like amphibians but as yet no frog-like amphibians.

It was at this time, possibly in these swamp-forests, that one of the most significant events in the evolution of vertebrates took place: the origin of reptiles. There are only about a dozen types of reptile known from the Upper Carboniferous, but they are sufficient to demonstrate that the

Dolorthoceras sociale, *Ordovician of the United States. It is the pluricellular shell of a cephalopod of the group* Orthoceras, *the ancestors of modern nautili, characterized by a straight shell*

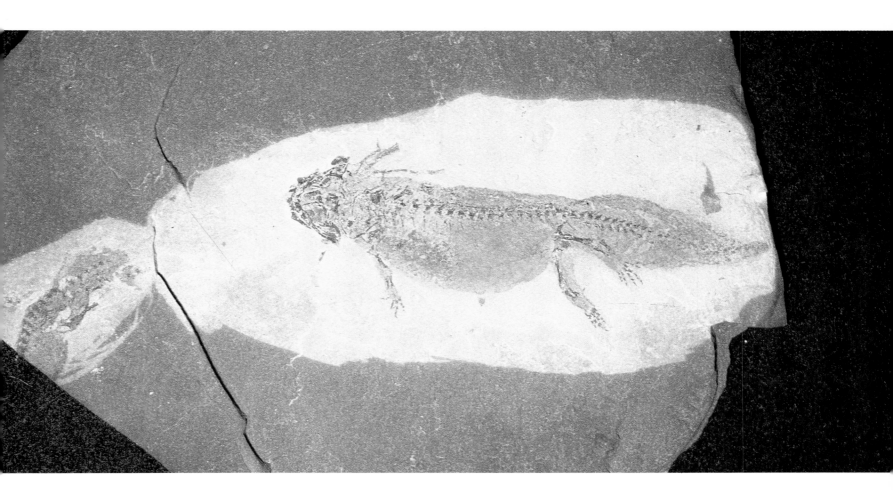

Skeleton and print of the body of a Branchiosaurus, *a small temnospondyl amphibian of the Lower Permian, Germany. This animal shows a quite unusual characteristic: it seems to have retained its external gills, typical of the larval stage, even in adulthood. If so, it would represent the product of an evolutionary process called neoteny, which consists of an accelerated development of sexual organs by comparison with the rest of the body*

"invention" of the amniotic egg had taken place in the Lower Carboniferous or earlier, and that fully terrestrial vertebrates existed. By the Lower Permian, reptiles and amphibians occur together in roughly equal diversity and by the Upper Permian, the reptiles predominate in terrestrial faunas.

Most Carboniferous reptiles are known from single specimens in swamp-pool faunas, and these probably represent terrestrial forms which have ended up in the pools by accident. Most belong to the most primitive order of the class, the Captorhinomorpha, from which other reptile groups are believed to have originated. They were small insectivores, less than 80 cm long in the Carboniferous and bore a superficial resemblance to lizards. In the succeeding Permian epoch the climate was either more seasonal or consistently drier and less hospitable than in the Carboniferous and it was then that many types of reptile appear in the fossil record, commencing the diversification which led them to fill so many terrestrial, aquatic and aerial niches throughout the Mesozoic. Not ony their skeletons but also their footprints testify to their abundance.

As discussed earlier, two different geographical areas were formed during the Permian, each characterized by a different climate, flora and fauna. At the beginning of the period, for instance, the northern continents were covered by a luxuriant flora very similar to that of the Carboniferous, followed by a drier climate and the development of the first conifers.

Mazon Creek

The three fossiliferous ironstone nodules found at Mazon Creek and here illustrated give an idea of the large variety of organisms preserved within the deposit and of their perfect conditions. Right: a nodule containing the leaves of a Neuropteris. Left: a specimen of Achistrum, a rare holothurian. Below: a specimen of the annelid Nereis, the delicate body of which rarely fossilizes satisfactorily

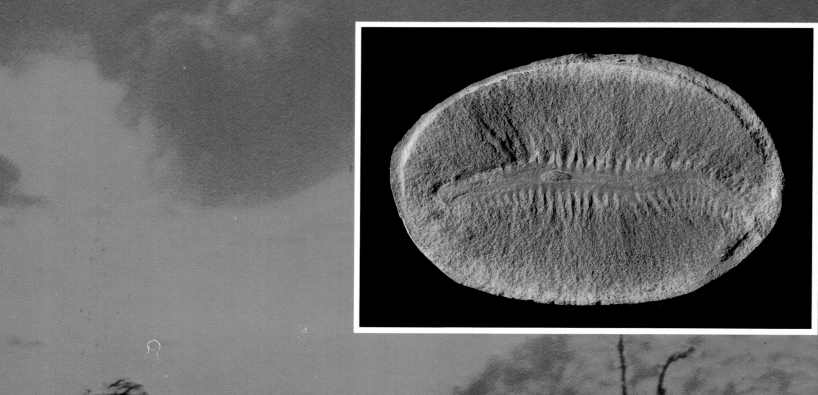

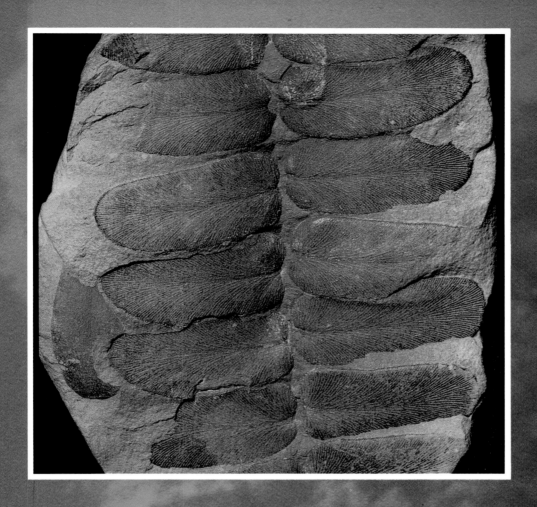

A particularly important role in the reconstruction of the history of life is played by certain palaeontological deposits which are particularly rich in fossils. These deposits formed under conditions which were especially favourable to the fossilization of organisms; they thus provide a window on the past, through which we can get to know animals and plants which, under different circumstances, would probably not have fossilized. One of the most famous among Palaeozoic deposits is the one at Mazon Creek, Illinois (United States of America). It consists of reddish ironstone nodules containing various types of continental plants as well as of marine and freshwater animals, which had deposited in a deltaic environment during the Carboniferous.

The two fossilized branchiopods in these photographs belonged to two different genera of the same family: Paraspirifer *of the Middle Devonian of the United States (above) and* Cyrtospirifer *of the Belgian Devonian (below). Their shells show similar general characteristics but a very different shape. The large variety of shapes which characterized the brachiopods of the Palaeozoic and Mesozoic has enabled scientists to use these animals as zone-fossils, thus determining the relative age of rocks, and to carry out correlations between sediments in different parts of the world*

The southern continents, on the other hand, were characterized throughout the period by a drier climate and less varied flora, the so-called *Glossopteris* flora, after the predominant pteridosperms. This led to a clear differentiation in the fauna and the possibility of a different evolution of the vertebrates. The humid climate of the northern continents, at the beginning of the period, still favoured the diffusion of labyrinthodont amphibians, which reached large dimensions: among them, *Seymouria*, the form showing characteristics of both amphibians and reptiles and once regarded as a classic example of the link between these two classes. Labyrinthodont amphibians had become less frequent, but reptiles were flourishing: captorhinomorphs and synapsids being abundant and diverse, the latter being the group which in the course of the following era, would give rise to the first small and primitive mammals. The Permian reptiles included many unusual and strange animals including *Dimetrodon* and *Edaphosaurus* (which had on their backs a thermoregulatory organ in the shape of a dorsal sail), and the mesosaurs (small freshwater reptiles whose presence in the Permian lakes of South Africa and South America is one of the palaeontological proofs of the union, in this period, between the two continents).

The bed of the salty lake of Timoun, Algeria, which evaporated a few hundred years ago. Many salt deposits of the Permian probably originated as a consequence of the similar evaporation of marine waters in desert areas

The Mesozoic Era

The Mesozoic Era

The Mesozoic began 230 million years ago, comprised three periods (Triassic, Jurassic and Cretaceous) and lasted 165 million years in all. Its biological and geological characteristics have progressed from those of the preceding era: in the course of the million years of its history, the Mesozoic witnessed the changes which caused the fauna, the flora and the geography of the planet to adopt an appearance very close to the present one.
The changes which took place were many. Geologically, Pangaea split into smaller continental masses, roughly corresponding to the present ones. Biological innovations were also considerable: the continental fauna was dominated by reptiles, particularly the dinosaurs, while reptiles of large dimensions lived in the seas, and flying reptiles dotted the skies. The foundations were also laid for the development of mammals and birds, the animals which predominate today. They appeared during the Triassic and Jurassic respectively.
The end of the Mesozoic was marked by a very important event: the extinction of large numbers of animal groups, among which were many of the predominant reptiles. Their extinction paved the way for the development of mammals and birds, and, consequently, for a renewal in the modern sense of the Earth's biology.

This rock of the German Triassic represents a fossilized sea bed of 200 million years ago, complete with the organisms which lived in it: some lamellibranchs together with some crinoids

The Mesozoic, or Secondary era, lasted 165 million years: a very long period of time stretching from the end of the Palaeozoic, some 230 million years ago, to the beginning of the Tertiary era 65 million years ago.

As the name implies, the Mesozoic was a middle era, a transition period between the archaic Palaeozoic world, which was so different both biologically and geologically from the present one, and the more modern Cenozoic, a period when the geographical and biological modifications laid the foundations of the modern world.

However transitional, the Mesozoic had its own unmistakeable characteristics. Its deposits, younger than the Palaeozoic ones, have yielded much better preserved biological documentation, and the rocks themselves, widespread throughout the world, have allowed detailed geographical, environmental and biological reconstructions. The biological and geological events which took place during the Mesozoic are therefore well known. The distribution of lands and seas, the diversity of fauna and flora, the appearance and disappearance of organisms in the course of evolution, have all been reconstructed with much greater precision than has so far been possible in the case of the much older Palaeozoic.

Like the Palaeozoic, the Mesozoic is also divided into smaller intervals of time; its three periods are in turn divided into even smaller ages. The first period of the era is the Triassic, so called because German geologists, who first defined it, attributed to it three successive types of rock: variegated sandstones (or Buntsandstein), shell-bearing limestones (or Muschelkalk) and variegated marls (or Keuper). The Triassic is thought to have lasted thirty million years. It was followed by the Jurassic, named after the Jura mountain range, where rocks of this age are very well represented. The Jurassic lasted sixty million years and was followed by the Cretaceous, the name being derived from the Latin *creta*, the Roman word for chalk which is widespread in France, Belgium, Great Britain and Germany. The Cretaceous lasted seventy-five million years and its end, sixty-five million years ago, was marked by an event the causes of which are still unknown: the simultaneous extinction of many groups of animals, among them the dinosaurs and the large marine reptiles, which were predominant throughout the Mesozoic.

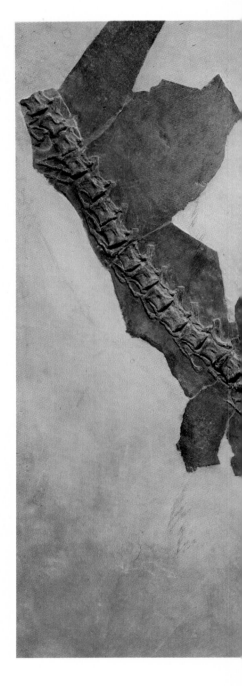

The Triassic

The beginning of the Triassic saw little modification with respect to the preceding Permian: the huge single continental mass was still there, with its eastern shore indented by the gulf of Tethys. The vast dimensions of this land favoured a continental type of climate. Triassic continental sediments, sandstones of aeolian origin mixed with marls and clays, seem to point to an arid climate, occasionally desert; sandy areas must have been extensive — they appear today as red sandstone with crossed stratification. Long periods of particularly dry climate alternated with more humid ones which allowed the formation of swamps and marshes and abundantly differentiated flora and fauna.

Askeptosaurus italicus, *discovered in the black Triassic schists of the Besano site, near Varese (Italy). It is a large reptile which lived in the marine waters near the coast, and belonged to the order Eosuchia, the ancestors of modern lizards. This specimen is almost two metres long*

During the Triassic, the sea level often changed considerably. Towards the middle of the period the sea covered part of the northern lands, forming shallow continental seas where life developed rapidly and which affected the climate, making it more suited to the development of life on the land.

It seems that the Triassic climate was warmer than the present one. This is proved by various factors, such as desert continental deposits, and marine deposits rich in coral reefs. Most corals need warm, tropical waters, with temperatures no lower than 20° C. This must have been the average temperature of water during the Triassic, because the reefs are widespread

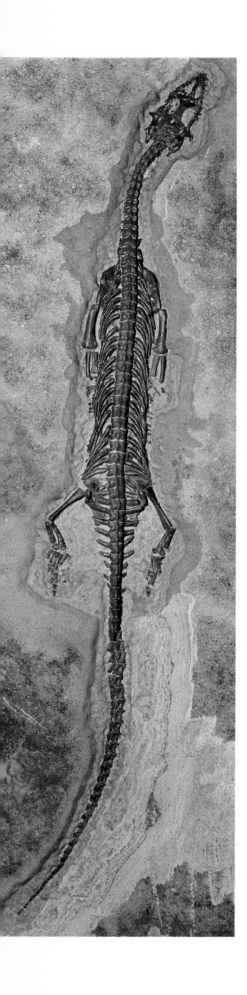

and point to a tropical environment. Some of these reefs, now huge mountain ranges, are well-known: first among them the Dolomites, which represent the remains of cliffs formed in shallow seas, the bottom of which was progressively sinking.

During the Triassic, although dry periods alternated with humid ones, the overall climate was certainly drier than in the preceding period. This seems to have been proved by the vegetation, which is well represented by fossil remains. The flora of the Triassic was quite different from the flora of the Palaeozoic; the characteristic elements of the Permian/Carboniferous periods disappear (all those tree ferns, lycopods, equiseta, and gigantic pteridosperms) and their place is taken by conifers, which covered vast areas with huge forests and continued to expand throughout the period. From a botanical point of view, one might say that the Triassic was the period of conifers, these plants suited to dry habitats, which became a predominant aspect of the flora and are still extremely numerous in their fossilized state. They were very close to the modern Araucaria; they could be 30 metres tall and the diameter of their trunk could reach over 2 metres: one was measured in the well-known petrified forest of Arizona, where huge trunks have been preserved transformed into multicoloured opal. Among the best known conifers found in European sediments are *Walchia* and *Voltzia*, of which only fronds are known.

Beside the conifers, other plants appeared in the Triassic which were destined to become typical of the Mesozoic. These were the first cycadales, similar in shape to today's ferns but extremely primitive. Today, they are represented by a few genera, among them the *Cycas*, often cultivated as ornamental plants in Mediterranean gardens.

Warm seas, analogous to modern tropical waters, greatly enhanced aquatic life, since tropical habitats favour form differentiation. This differentiation was encouraged, in continental environments, by the climatic alternations. On the one hand, the huge coniferous forests and the desert regions allowed the development of animals adapted to arid conditions; on the other, swamps and marshy lands were more suited to the life of amphibians and reptiles.

Life forms during the Triassic were therefore particularly diverse. Very important groups developed in the seas: corals were widespread, ammonites rapidly evolved (they had first appeared during the Devonian and had been subjected to noticeable reduction in numbers at the end of the Palaeozoic). These cephalopods became numerous enough to be considered excellent guides as to the relative age of the various geological strata. The first belemnites developed alongside them. Together with modern-type crustaceans, molluscs and echinoderms became widespread.

Certain sediments, such as those at Besano in Lombardy (Italy), have proved that marine reptiles were indeed numerous, and one could say that during the Triassic, reptiles reverted to aquatic habitats, spreading through many seas and developing various forms and ways of life. The reptilian group which became most specialized to a marine existence was the Ichthyosauria. Their limbs were modified as paired fins, their long tail also possessed a fin, and they had a hydrodynamic body-shape similar to those of sharks or dolphins. The first ichthyosaurs appeared in the Triassic.

This marine rock of the German Middle Triassic consists almost entirely of the fragments of crinoids. Similar organisms have substantially contributed, throughout the eras, to the formation of such rocks, called organogenic

On the previous page: *the skeleton of a* Pachypleurosaurus edwardsi *from the Triassic of Monte San Giorgio (Italy). This was a small reptile belonging to the Nothosauria, the ancestors of the large Jurassic plesiosaurs. Nothosaurs and plesiosaurs were probably related to the modern lepidosaurs*

They were the mixosaurs, already adapted to aquatic life, although the tail was not as fish-like as in later forms. The ichthyosaurs reached their greatest diversity in the Lower Jurassic and several different species have been found together in the Liassic at Lyme Regis in Dorset (England). Their fish-like body-shape was only possible because (unlike sea-turtles) they had overcome the problem of laying shelled eggs on land by becoming ovoviviparous — giving birth to live young in the water.

Another group of aquatically adapted reptiles was the Placodontia. The placodonts were exclusively Triassic forms up to 1.5 m in length, and characterized by the possession of sets of plate-like teeth on the jaw and the palate for crushing and grinding the shells of molluscs and crustaceans.

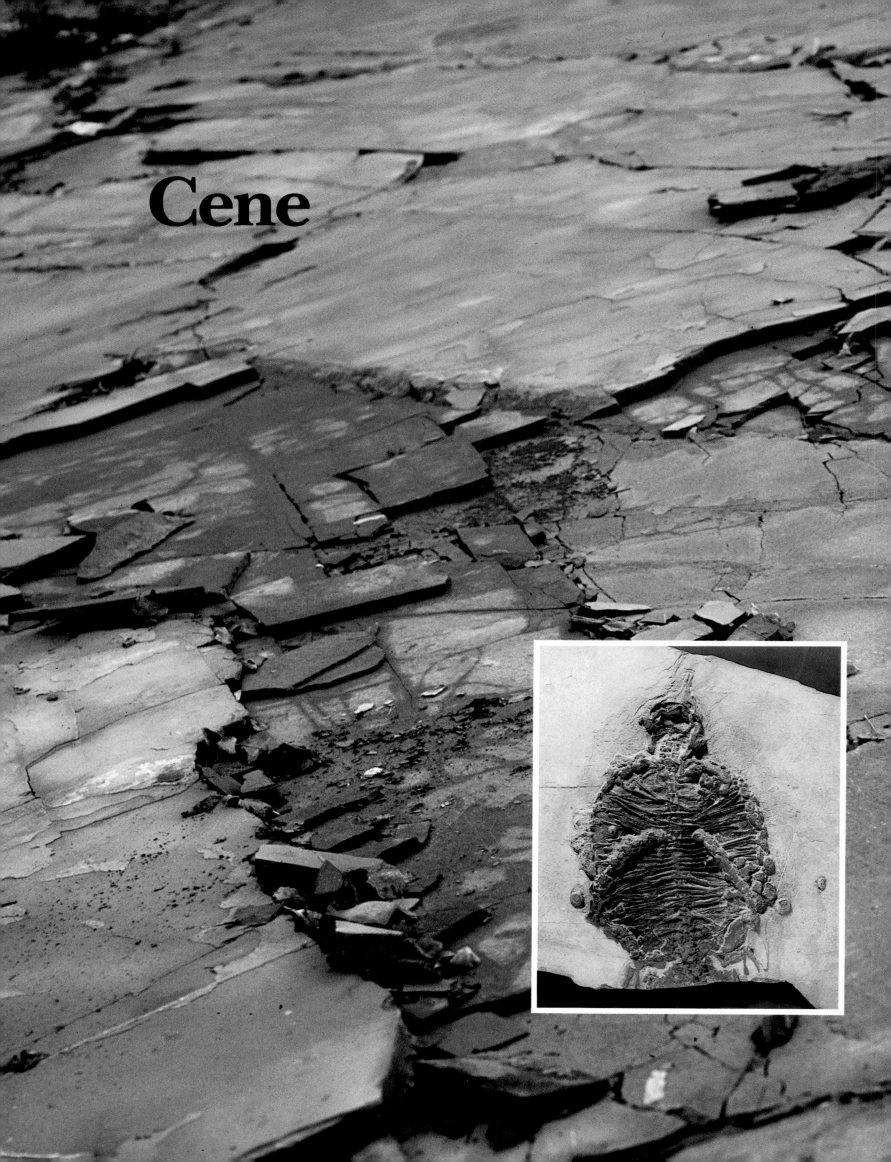

Cene

A very interesting deposit was discovered in Italy a few years ago. In a cave near the township of Cene, as well as in other parts of the area around Bergamo, a rock formation dating from the Upper Triassic has been identified. The layers of the formation, probably originating in a lagoon, were extremely rich in fossils: crustaceans, fishes, molluscs, reptiles and plant fragments, they all enable us to explore the composition of life some 200 million years ago. The fossils found at this site are still being studied, and research in situ continues to unveil new specimens.

Right: *a lepidosaurian reptile with huge claws and a long spine on the tip of its tail, an unusual feature in a reptile. This Drepanosaurus unguicaudatus is undoubtedly one of the most interesting fossils to be found in the Triassic rock of Bergamo*
Left: *the complete skeleton of a placodontian reptile, fossilized on its back*

A fish from the Triassic deposits of Cene (Italy). Its state of preservation is perfect not only for the scales, which were rather tough, but also for the delicate details of the pectoral fins and the caudal fin

Eudimorphodon ranzi *is the oldest known flying reptile. A few specimens have been found, only in the Triassic rocks of Bergamo (Italy). The extremely well preserved type-specimen illustrated is the one upon which the description of the species is based*

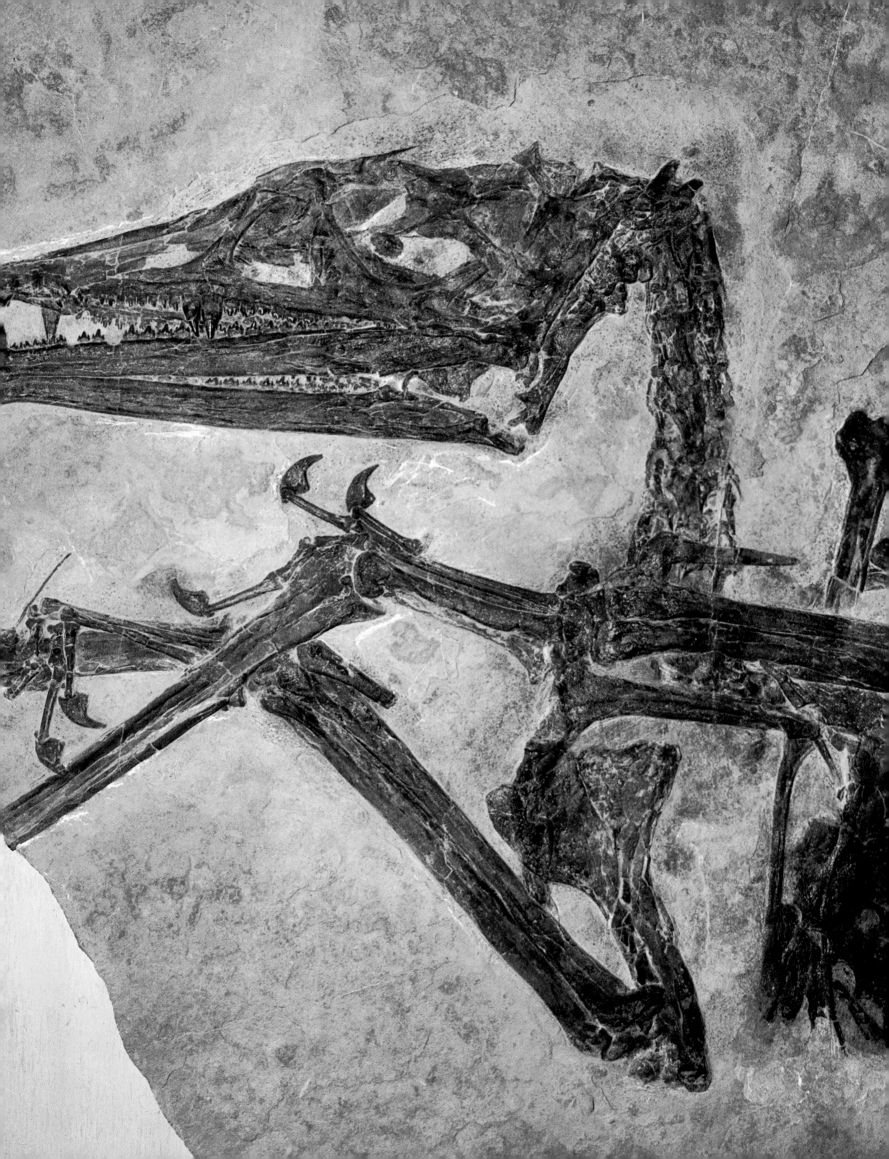

Some were clad in bony armour and bore a superficial resemblace to turtles. A third group of exclusively Triassic marine reptiles was the Nothosauria. They appear to have been semi-aquatic lizard-shaped forms, capable of walking on beaches and reefs as well as swimming and feeding on fishes. They were abundant in Triassic lagoon deposits and it is believed that the larger plesiosaurs and pliosaurs of the later Mesozoic were their descendants. The plesiosaurs appear at the base of the Jurassic, replacing the nothosaurs and were mostly small-headed long-necked aquatic specialists with paddle-like fins.

Turning to the continental animals of the Triassic, we may first mention the labyrinthodont amphibians. They were still abundant, sometimes large, crocodile-like animals, like *Mastodonsaurus* of the European Triassic which had a skull over 1.5 m long.

The first lepidosaurs, the ancestors of the modern lizards and snakes, became abundant in the Triassic. Some lived in coastal waters but most were strictly terrestrial small insectivores. Among the first true lizards appearing during the period were several strange gliding forms which moved by use of a membrane of skin stretched along trunk ribs which were long and extended out from either side of the body.

During the Triassic, the dry lands were densely populated. Reptiles included the first and most primitive tortoises, the genus *Proganochelys*; the

Below left: Laevaptychus, *which could reach over 26 cm in width. It is, in fact, the operculum of a gigantic ammonite.* Right: *an ammonite of the genus* Amaltheus, *from the German Jurassic, still preserving fragments of the original shell.* Below right: *a small unrolled ammonite of the genus* Crioceras *from the alpine Cretaceous*

thecodonts, ancestors of dinosaurs and crocodiles, the footprints of which are very common in Triassic rocks; the first dinosaurs, found in America, South Africa and Europe; the earliest flying reptiles — the pterosaurs, found in the deposits of Cene, near Bergamo (Italy); and, finally, the therapsids or mammal-like reptiles.

The latter were extremely widespread throughout the period, particularly in the southern continents. They are the group of reptiles from which mammals descended and all of them show a tendency to adopt mammal characteristics in the structure of skull and limbs. Many show a close approach to a mammalian form but only one group of very small forms did achieve it, thus originating, during the Triassic, the oldest and most primitive of known mammals. Therapsids have been discovered in large quantities, but we should at least mention the genus *Lystrosaurus*, fossilized remains of which have been found in South Africa, India, China, Russia and Antarctica, thus proving again that these continents were once one mass. The most important events to happen in the Triassic, as far as evolution is concerned, were two: the appearance of the angiosperms in the vegetable world, as documented by the findings of fossilized palms in the Middle Triassic of Colorado (United States of America); and the appearance of mammals, a few fossilized remains of which have surfaced in many parts of the world.

Left: *a calcareous sponge of the English Cretaceous,* Raphidonema farringdonense. *A fossilized sponge in such good condition is rarely found. These organisms usually disintegrate after death and only their spicules (the elements of the internal supportive skeleton) are preserved in the sediments*

Right: *The skeleton of a small sea turtle dating from the Cretaceous. This fossil, rather rare in Italian deposits of this period, was found in the Bergamo area*

The Triassic was, therefore, a period of great evolutionary innovation. It was also rather unfortunate as far as certain animal groups are concerned. A biological crisis hit continental animals at the end of the period. Many groups became extinct, among them the placodonts, nothosaurs, thecodonts and most labyrinthodont amphibians and therapsids. This was not, however, an entirely negative phenomenon, as it paved the way for, and opened new ecological niches to colonizing species, thus allowing the survivors to evolve more rapidly. The biological crisis of the Triassic allowed the fauna to renew itself, as can be easily seen in the much different Jurassic fauna.

The Jurassic

The continental masses of the Jurassic (started 195 million years ago) appear to have maintained the same basic positions they had in the preceding period. According to some geologists, the huge, single continental mass of Pangaea still existed.
Jurassic sediments and fossils show that the period was characterized by great uniformity in climatic conditions and in the distribution of land and sea. The climate seems to have remained very warm and rather humid throughout, allowing the formation of vast tropical forests over all the dry lands, with the exception of a few limited areas where almost desert conditions prevailed. Overall, the Jurassic was a luxuriant period: forests, wide rivers, vast marshlands and huge estuaries encouraged a varied and abundant life, as well as the development of all the many forms which make Jurassic fauna one of the most interesting of the geological past.
Jurassic flora was luxuriant: it witnessed a partial decline of conifers (still widespread in some areas) accompanied by the development of other plants. Vast areas were covered by cycadales, bennettitales and gingkoales — the most widespread of Jurassic plants — while angiosperms still played a subordinate role. Within these forests the dinosaurs evolved; these reptiles, divided by scientists into the two groups Ornithischia and Saurischia, became the real lords of dry lands, with forms of all dimensions and adapted to all environments. They included the gigantic quadrupeds, such as *Brachiosaurus* and *Diplodocus*, adapted (like giraffes) to browsing on the canopies of trees; the heavily armoured stegosaurs; the carnivorous dinosaurs of the Carnosauria, such as the predator *Allosaurus*; and the smaller, swift dinosaurs, predators of smaller animals and eggs. Life next to such neighbours must have proved rather difficult for the other land vertebrates, as is amply proved by the reduction in their numbers. The other Jurassic land vertebrates were not as numerous as dinosaurs and were usually much smaller (like the mammals and many reptiles), or evolved in different environments from those inhabited by the dinosaurs: they took to the sea and the air. That is why this period saw the great success of flying reptiles, while the seas became populated by large and curious forms.
Marine reptiles are documented by the fossils found in such well-known deposits as those of Solnhofen and Holzmaden in Germany, and Lyme

Complete skeleton and detail of the skull of a specimen of Liopleurodon ferox *of the English Jurassic. It is a plesiosaur of considerable size, belonging to the pliosaur group which was characterized by a short neck and large skull*

Gadoufaouà

In the Teneré desert, some 200 km east of the town of Agades in the Niger, one of the largest known deposits of dinosaur fossils was found. The site consists of a strip of land about 1 km wide and over 150 km long, with clays and sandstones of various colours from red to green, containing an enormous quantity of fossilized remains. These sediments were deposited during the Cretaceous, in a marshy area covered in vegetation in which crocodiles, turtles and a large variety of dinosaurs lived. This particular site, called Gadoufaouà, was only recently discovered and has been only partly explored. It often happens that the wind, shifting the sands of the dunes, uncovers whole dinosaur skeletons.

Some of the fossils found in the Gadoufaouà site. Right: the marks left by the three toes of a bipedal carnivorous dinosaur. Below left: the sharp tooth of a fossilized crocodile. Below right: the vertebral column of a large dinosaur emerging from the desert sands

Regis in Great Britain. Marine reptiles of the period included the large ichthyosaurs, direct descendants of the Triassic mixosaurs, the long-necked plesiosaurs and the short-necked, long-headed pliosaurs, both descendants of the nothosaurs and more specialized swimmers than the latter. Side by side with these vertebrates, some of which could be up to 10 m long, lived all sorts of invertebrates: a large variety rendered possible by the tropical conditions of the environment. Ammonites were particularly numerous and represented by many different genera; and so were belemnites, similar to modern squids. The remains of these cephalopods are now so plentiful that they can always be used as a guide to the various chronological layers of the period.

However, any attempt to establish which was the most important evolutionary event to take place during the Jurassic would probably highlight the appearance of the earliest birds as shown by *Archaeopteryx lithographica*, a few specimens of which have surfaced in the Upper Jurassic limestones near Solnhofen and Eichstatt, in Bavaria (Germany). This primitive flying vertebrate, endowed with real feathers, shows a mixture of the characteristics of reptiles and birds and demonstrates not only that the class of birds had already developed during the Upper Jurassic, but also that these flying vertebrates were the descendants of a group of agile, carnivorous dinosaurs.

The Cretaceous

While the Jurassic was still characterized by a flora and fauna different from today's, the Cretaceous was, in many respects, a more modern period. This also applies to its geography; that is to say, to the disposition of the continental masses. What actually happened during the Cretaceous is extremely important: the huge single continental mass began to split. First the central Atlantic opened up, then the South Atlantic until, towards the end of the period, the continental masses assumed the shape we know today, with only two exceptions: India was still an island in the middle of the Indian Ocean, and Australia and Antarctica were still joined together.

Such a geographical change was bound to affect considerably both the structure and the composition of the fauna; the period was thoroughly conditioned by these geological events, particularly towards its end, when the most serious biological crisis ever seen on Earth took place.

The transition between a single continental mass and a conglomerate of continents divided by deep oceans brought about considerable environmental changes. There is evidence that during the Cretaceous there existed a variety of climates never seen before, and that for the first time the seasonal cycles appeared.

A climate based on seasons obviously produced quite a change in the vegetation: the plants which had flourished in the Jurassic lost their foothold to the angiosperms, and the flora began to look very modern, with plants such as the oak, the magnolia, the plane and the laurel, which are still important elements of the modern botanical heritage.

The skeleton of an Allosaurus fragilis *on show in the Museum of Natural History in Milan. The allosaur was a carnivorous dinosaur 7 m long, which lived in North America during the Jurassic. It was a bipedal predator, which could move swiftly and was endowed with a formidable set of teeth*

The Cretaceous climate was generally warmer than the present one, judging by the plant fossils found in areas like Greenland, where the present climatic conditions are uncomfortable to say the least.
There was a considerable turnover in herbivorous dinosaurs. Flying reptiles evolved considerably, often reaching the gigantic dimensions of over 10 m of wing-span. In the seas the huge mosasaur appeared — a sort of long serpent with small fins, related to today's lizards. The last of the plesiosaurs were reduced in numbers, albeit some of them reached enormous proportions, and so were the ichthyosaurs. On the other hand, birds were more numerous than during the Jurassic; various skeletons have been found in Cretaceous sediments which show primitive birds, some of them still endowed with teeth. In the seas, ammonites and belemnites lived in large numbers, side by side with the large reptiles and all types of fishes and invertebrates.
Towards the end of the Cretaceous, all these animals were affected by the biological crisis already mentioned. Over a short period of time many groups of organisms became extinct: all the dinosaurs, the large marine reptiles, the ammonites, the belemnites (almost completely), and many other less striking groups. The overall taxonomic diversity was suddenly cut down to a very low level, particularly in the sea.
For many years the reasons for such a vast biological crisis and so widespread an extinction were considered an almost insoluble problem: palaeontologists could not explain the disappearance of animals so different as, for instance, ammonites and dinosaurs, without resorting to catastrophic events such as the explosion of a nova or the impact of a huge meteorite. The real reason of the biological crisis, however, is much more natural, as it is linked with environmental changes which took place during the Cretaceous as a result of the breaking up and the movements of the continental masses.
Geographical variations such as these deeply affected the climate; this in turn affected the fauna, leading to the extinction of the most widespread among the animal groups: those which were better adapted to the prevailing climatic conditions and were therefore incapable of adapting quickly to new conditions in order to survive. The biological crisis which took place at the end of the Cretaceous was, once again, not entirely negative; it paved the way for a general modernization of the fauna over the whole globe, a process which started at the beginning of the following era, the Cenozoic.

Tyrannosaurus rex, *the largest known carnivorous dinosaur, about 10 m long, was even larger and more powerful than the allosaur. It was also a carnivorous biped, adapted to the life of the Cretaceous forests, quick and aggressive, with extremely powerful teeth, as shown in the photograph*

Osteno

One of the most interesting of all Jurassic deposits is the one near the village of Osteno, on Lake Lugano (Switzerland). It consists of a series of calcarosiliceous strata where marine organisms have been found in extremely good state of preservation: they still retain all their anatomical details and often even traces of the soft parts of their bodies. There are fishes and crustaceans, annelids, echinoderms and molluscs which convey a clear idea of the fauna living in Jurassic seas. The terrestrial plants also found on the site give us excellent information on the plants which covered the Jurassic dry land.

Three of the best specimens found in the Osteno deposit. **Left:** *a completely preserved fish.* **Centre:** *a crustacean of the genus* **Aeger** *with its delicate legs and antennae.* **Right:** *a crustacean of the genus* **Coleia,** *now extinct*

Solnhofen

The most famous Jurassic deposit, perhaps even the most famous deposit in the world, is that of Solnhofen and Eichstatt, in Bavaria (Germany). The stratified yellow limestones of the Upper Jurassic which come to the surface in the countryside between the two towns contain large numbers of perfectly fossilized organisms. The rocks are what is left of an ancient lagoon, where waters were calm and poor in oxygen, devoid of destructive agents, particularly biological ones such as bacteria and predators. All these factors meant that the organisms born by the waves inside the lagoon could completely fossilize after death. This deposit therefore has preserved traces of the soft parts as well as delicate anatomical elements, without which no important discovery would have been possible.

Mesolimulus walchii, found in the Jurassic limestone of Solnhofen. This organism was analogous to the modern king-crab Limulus, the last representative of merostomous arthropods. Having first appeared in the Triassic, these organisms have survived almost unchanged for over 200 million years

A crustacean of the genus Aeger of the Solnhofen Upper Jurassic, which still retains all the anatomical characteristics of the skeleton. Its cephalic appendages are clearly visible, including the delicate antennae

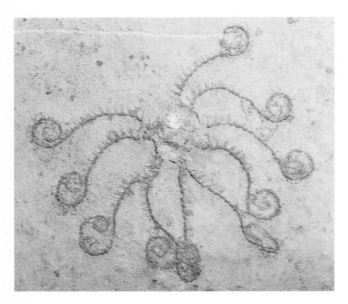

Left: *the fossil most typical of the Solnhofen and Eichstatt site, and probably the most common, is* Saccocoma pectinata, *an echinoderm belonging to the group of the crinoids. It can be found in large quantities on the slabs of yellow limestone of the deposit. It is often mineralized and not clearly visible, occasionally it is perfectly preserved, like the one in the photograph*

Right: *An insect found at Solnhofen,* Cymatophlebia longialata. *The particularly fine grain of the rock encasing it has preserved even the most delicate of its structures, such as the membrane of its wings*

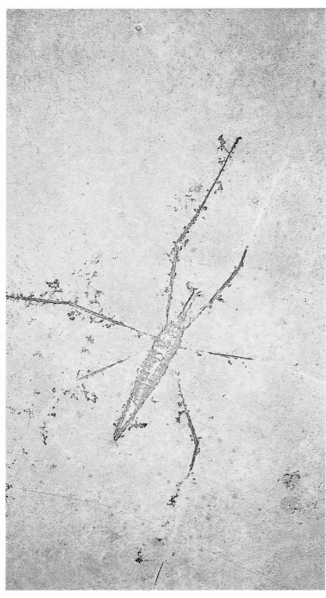

Left: *Another insect found in the Jurassic sediment of Solnhofen:* Stenarthron zitteli. *Insects abound in this deposit. They probably fell, during flight, into the waters of the lagoon, drifted to the bottom and were preserved together with the marine organisms*

The fish in the photograph belonged to the species Mesodon macrocephalus, *a reef-inhabiting creature with teeth suited to grind the corals and other calcareous organisms. This specimen probably lived on the outer side of the lagoon reef, and fell into the lagoon after death*

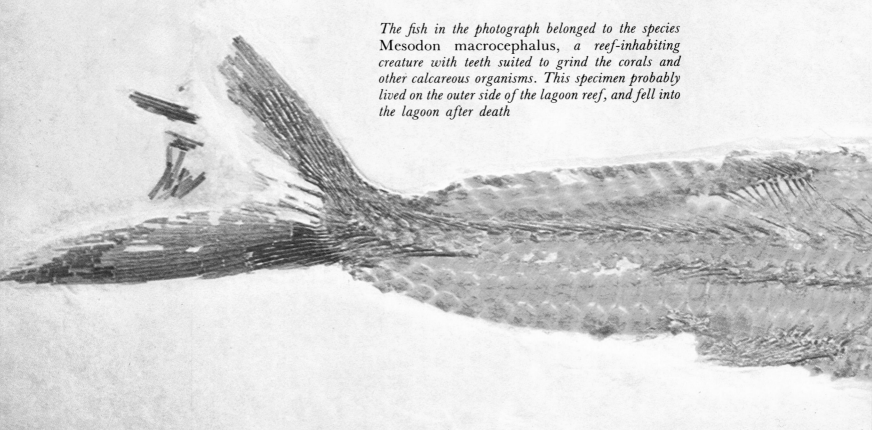

Fishes are naturally very numerous in the Solnhofen deposit. Many may have lived outside the lagoon and been thrown into it by the waves which broke over the reef separating the lagoon itself from the open sea. Once in the lagoon, animals would have been killed by the environmental conditions, such as high salinity and low oxygenation. The specimen in the illustration belonged to the species Aspidorhynchus acutirostris

The yellow limestones of Solnhofen and Eichstatt have yielded numerous remains of pterosaurs, the flying reptiles which, in the Upper Jurassic, occupied the place now held by birds (then only at the beginning of their evolution). The fossilized pterosaurs found in the deposit belong to several species and demonstrate the large variety achieved by this group of reptiles. The specimen in the photograph is a Pterodactylus antiquus, *slightly larger than a sparrow*

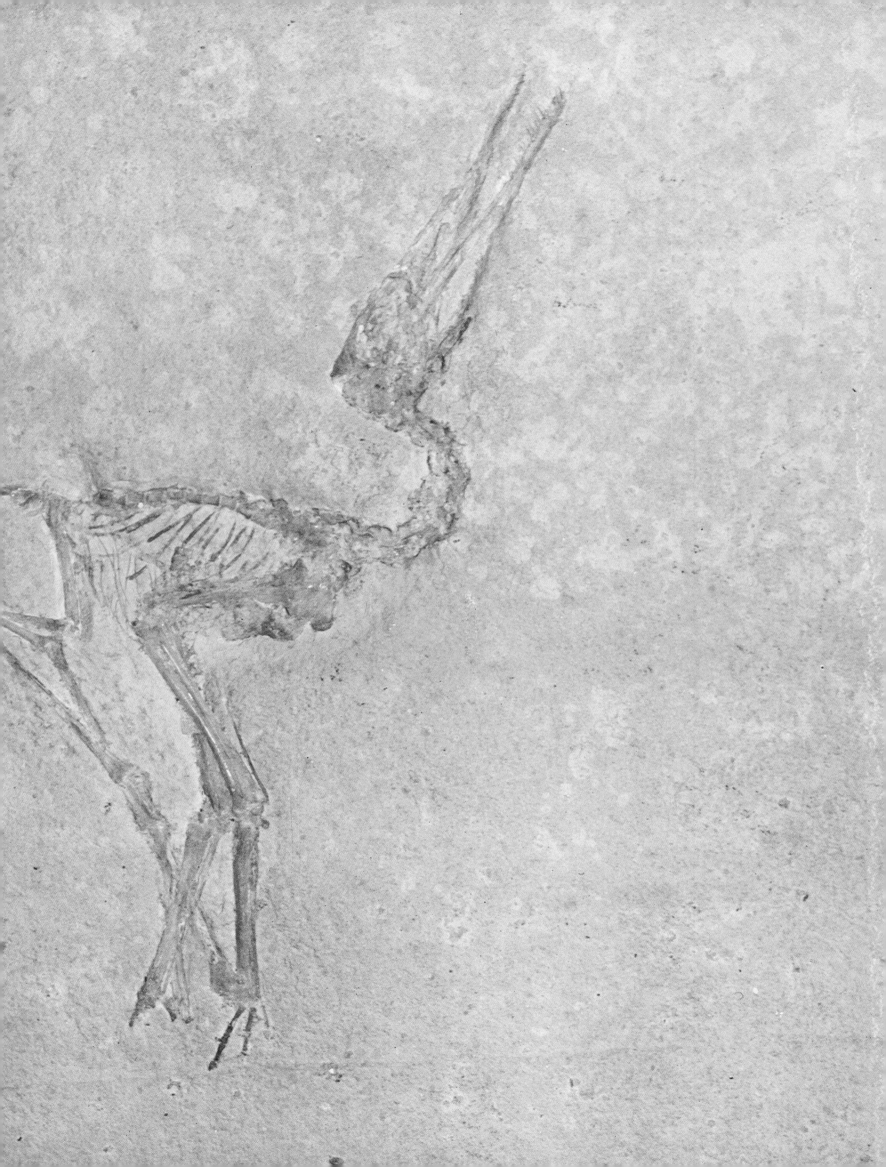

The Cenozoic Era

The Cenozoic Era

The Cenozoic began sixty-five million years ago and lasted sixty-three million years. During this time the geological and biological evolution of the Earth progressed to the threshold of history. Geography was affected by considerable changes and increasingly modern faunas followed one another, getting closer and closer to the appearance of the world as we know it today.
During the Cenozoic, an era of great changes, the continents continued to drift, assuming their modern shapes and forming the highest mountain ranges in the world, including the Alps, the Andes, the Himalayas, the Rocky Mountains. The fauna changed apace: at the beginning of the era several groups of mammals appeared which are today extinct, and formed an exceptional variety of species. In the course of the era, new groups replaced the old ones in a process of constant modernization. This uninterrupted sequence of forms has enabled palaeontologists to reconstruct, with the help of fossil remains, the history of most of the animal groups living today. This is the precious legacy left us by the Cenozoic: the chance to get to know the origins of the modern world.
The Cenozoic (also called the Tertiary era) means, literally, "the era of recent life" because both the fauna and the flora now become more and more modernized.

The marine fauna of the Cenozoic was very similar to the modern one. Ammonites, belemnites and other groups which had characterized the Mesozoic had vanished, and the organisms which had survived, although not identical, were certainly very similar to those of today. The two photographs (right) show two crustaceans of the Italian Eocene; one of them (inset) was found near Verona, the other in a calcareous nodule in the Ovada region

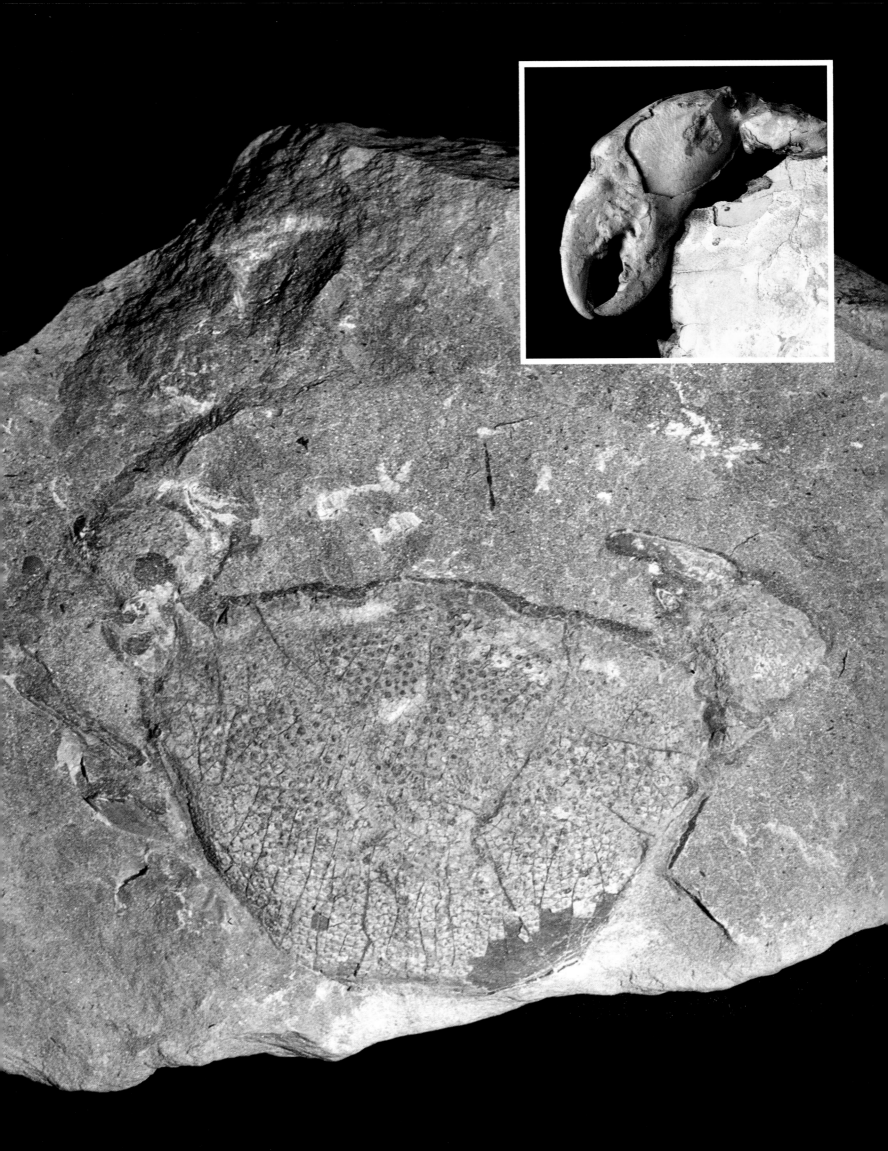

During the Cenozoic considerable geographical and environmental changes took place alongside important geological events. A series of successive modifications, involving the physical and structural part of the surface as well as the organic world, gradually changed the face of the planet until it reached its modern form: the continents as and where we know them today and a composition of both fauna and flora very similar to the present one.

Like the previous eras, the Cenozoic is also divided into shorter intervals and contains two periods: the Palaeogene and the Neogene. These are, in turn, subdivided into even shorter intervals called epochs. From the oldest to the most recent, they are called Palaeocene, Eocene, Oligocene, Miocene and Pliocene.

As we have said, the land masses and seas changed considerably during the Cenozoic, and finally reached their modern position. A geographical map of the Eocene conveys a strange feeling: the geography of fifty million years ago looks familiar notwithstanding a few curious details. North and South America were perfectly formed and Africa looked very much as it does today, except that the Arabian peninsula was welded to it. However, Africa and America were divided by a much narrower Atlantic Ocean than the modern one; Australia was still linked to Antarctica; and India still in the middle of the Indian Ocean.

Even more surprising was the sea which divided Africa from Eurasia —

A large block of white limestone from the French Miocene containing several sea-urchins. It is almost as if a piece of ancient sea-bed had been retrieved with the organisms which lived in it still in their original position

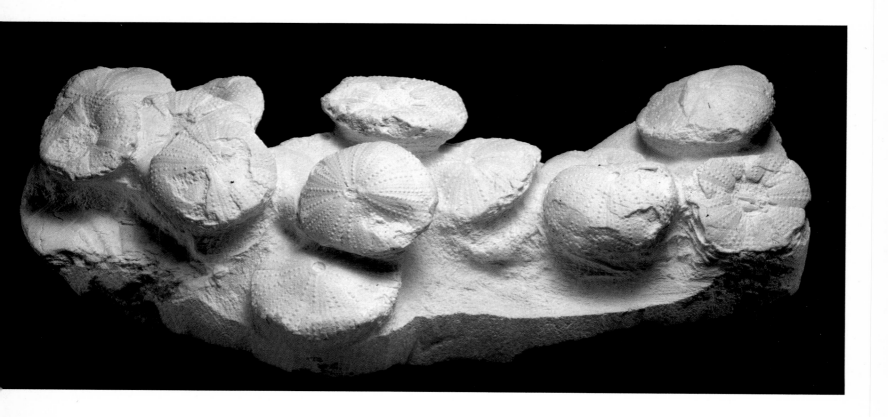

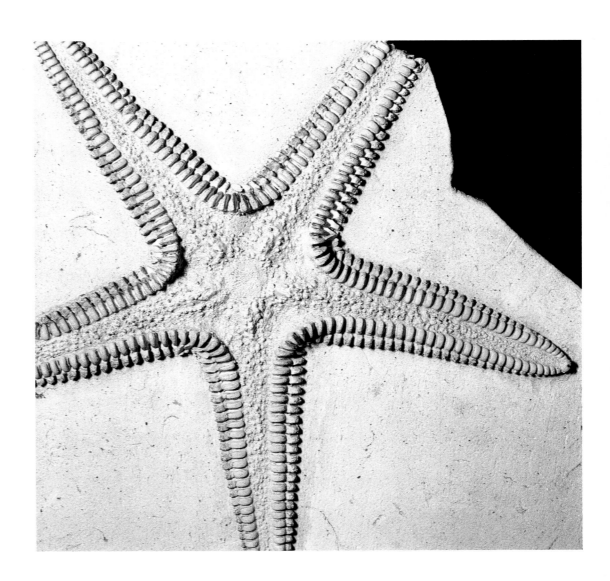

A starfish preserved within marine clays dating to the Pliocene and found at Castell'Arquato, near Piacenza (Italy). The skeleton of the echinoderm was not preserved, only its cast in clay

the Tethys — had vanished and had been replaced by a proto-Mediterranean from which Italy, most of Greece and Turkey were missing. This sea was open to the east and communicated with the Indian Ocean. Europe and Asia lacked the mountain ranges which characterize them today: the Pyrenees, the Alps and Apennines, the Anatolian mountains, the Caucasus and the Himalayas. In America, the Andes and the Rocky mountains were not completely formed. All these mountain ranges took shape during the Cenozoic: the European and Asiatic ones due to the compression of sediments caused by, respectively, Africa and Arabia approaching Europe, and India joining Asia; the American ones formed by a phenomenon connected with the Pacific sea-bed gradually sinking under the continent itself. As for Italy, apart from a few islands dating to the Triassic, it first appeared during the Cenozoic. The Eocene saw the formation of some large islands in what is today Tuscany and southern

Italy; during the Oligocene, these islands became more substantial and, as a consequence of the formation of the Alps, dry land began to appear in northern Italy. During the lower Miocene, the Apennines appeared out of the sea followed by more dry land corresponding to modern Liguria, parts of Tuscany and Latium, and towards the end of the period the peninsula was almost complete: the Alps and Apennines rose to their present height, a marine gulf covered the Po valley and the sea reached up to the eastern foothills of the Apennines, the latter situation altering only very slightly until the end of the era.

The fauna and flora of the Cenozoic presented, right from the beginning, certain modern features. Vegetation consisted mainly of angiosperms and, since the climate of many of the periods was warmer than at present, it was mainly of a tropical kind, thick with palms, the fossilized remains of which have been found in many places all over the world.

This warm, tropical climate was not constant throughout the era: climatic fluctuations caused considerable environmental changes. However, the fossilized remains of tropical fishes and gigantic palms which have been found in Italian deposits (particularly the famous site of Monte Bolca) and the presence of madreporarian corals, clearly point to a much warmer climate than the present one at the same latitude. This type of climate obviously favoured the formation of large forests, within which mammals evolved with prodigious speed.

Fossilized molluscs, often perfectly preserved except for loss of colour, abound in Cenozoic deposits. Italian Pliocene clays have yielded the four specimens pictured here. Left to right: they represent the genera Narona, Murex, Fusinus *and* Cymatium

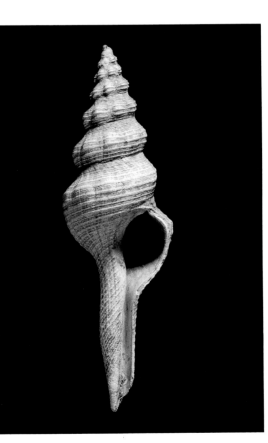

At the beginning of the era, once freed from competition from the dinosaurs (those reptiles which had occupied all ecological niches for large vertebrates and prevented other vertebrates from developing), mammals had become already fairly widespread. All orders known today appeared during the Palaeocene and Eocene, as well as other orders which became extinct before reaching the present era.

These extinct orders include the creodonts — small primitive carnivores; such archaic ungulates as the amblypods, best-known among them being *Uintatherium*, having a skull complete with tusks and bony horns; the desmostylans — aquatic mammals related to the elephants; the astrapotheres, the notoungulates and the litopterns — primitive ungulates exclusive to South America.

The evolution of many of these extinct groups and of the orders of mammals still alive today can be followed through the fossils dating from the various periods of the Cenozoic. To cite only two examples, we have been able to reconstruct the evolution of both horses and elephants. The ancestor of the former is the *Eohippus*, an Eocene animal no larger than a hare. The elephants descend from *Moeritherium*, the oldest known "elephant", without tusks or proboscis, which has been found in the Eocene lacustrine sediments of Fayum, in Egypt.

The Cenozoic saw an exceptional variety of mammals. Geological strata have yielded, in various parts of the world, the fossilized skeletons of strange animals, ranging from types lacking any particular specialization and from which several modern mammals may have descended, to highly specialized forms: ungulates with several impressive horns, gigantic aquatic mammals similar to modern hippopotamuses, and *Baluchitherium*, a proto-rhinoceros up to 5 m high.

Just as impressive was the variety of birds which developed in the course of this era. Among the most striking and unusual forms were *Diatryma*, a flightless Eocene bird, about 2 m tall and supplied with a very strong bill; and *Phorusrhacos* of the South American Miocene, a predator also incapable of flight and as tall as a man. These types of birds, adapted not to flight but to running swiftly on dry land, seem to indicate a certain tendency for birds to compete for territory with mammals at the beginning of the Cenozoic, when the dinosaurs had "recently" freed all ecological niches. The race to occupy the place left vacant by the dinosaurs was finally won by mammals, structurally more adaptable than birds.

Just as the disappearance of the large Mesozoic reptiles had allowed the development of continental mammals, the demise of marine reptiles enabled mammals to become adapted to marine life. Already at the beginning of the Eocene the cetaceans were there. They were primitive forms, their teeth indicating that they probably descended from a group of carnivores which had adapted to aquatic life. Apart from them, marine fauna remained the same throughout the era, very similar to today's. Belemnites and ammonites disappeared, modern-type cephalopods (such as squids and cuttlefish) appeared in quantities and the seas became very similar to modern tropical waters.

Reverting to life on dry land, it is now time to examine a group of mammals which will affect life on the planet for millions of years to come:

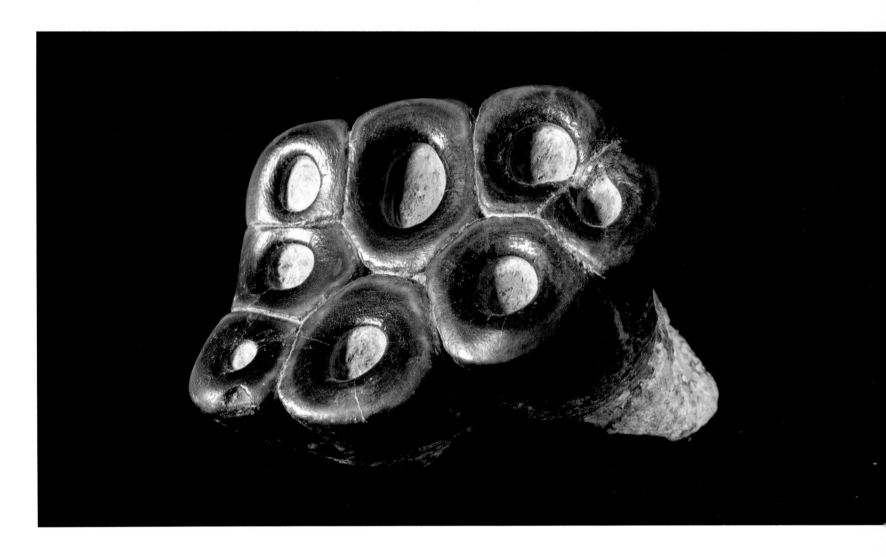

The tooth of a Desmostylus *of the American Miocene. This mammal belonged to the extinct order of the Desmostylia — peculiar animals adapted to aquatic habitats, as large as a hippopotamus and with characteristics similar to those of the subungulates and the elephants*

the primates. The Order Primates, to which the lemurs, monkeys, apes and man belong, have a very long evolutionary history which goes back to the beginning of the Cenozoic. Already during the Eocene, primates are present as a clearly distinct group. They evolved in the course of the era into increasingly specialized forms until the beginning of the Miocene. This period, at least fifteen million years ago, has yielded, in a Tuscan deposit, the *Oreopithecus*, a monkey endowed with a set of teeth and limb structure which make it probably one of the oldest representatives of the hominoid group. An even more interesting discovery from the late Miocene is *Ramapithecus*, a decidedly advanced primate, which some scientists regard as a hominid — that is, a representative of the group to which modern man belongs.

The dentition of *Ramapithecus* was not structurally different from that of other hominids of the Cenozoic, the australopithecines. The latter lived during the Pliocene, mainly in Africa. The oldest of these hominids so close to modern man is at least 3,700,000 years old.

Monte Bolca

Archaeocypoda veronensis, a *fossilized crab.* This specimen shows how the fossils found at the Monte Bolca site (Italy) still preserve almost all the anatomical details, including, as here, all their legs

The most famous Cenozoic deposits are those of Monte Bolca, near Verona. They consists of a series of calcareous strata which contain one of the most interesting faunas of the Lower Cenozoic; a fauna rich in fish, crustaceans, invertebrates, as well as terrestrial flora. This accumulation of fossils was probably caused by volcanic eruptions which poisoned the waters of a tranquil bay.

Two more fossils from Monte Bolca. Above: a plant preserved as if in a herbarium. Right: the most famous fossilized fish found in the deposit, Mene rhombea

The best known fossils from Monte Bolca are undoubtedly the fishes. A vast number of different species was found, ranging from the large sharks to the angelfishes with their delicate fins. The Bolca fossils are analogous to the fishes which today live in tropical sea waters, clearly indicating that a warm climate affected northern Italy at that period. The specimen in the photograph is a Sparnodus vulgaris. The whole skeleton is preserved, as well as parts of the body around it

The Quaternary Era

The Quaternary Era

The Quaternary only started two million years ago. Its beginning was marked by a noticeable drop in climatic temperature — a glaciation during which polar ice-caps covered areas which are now temperate, and almost continuous ice crusts formed on the main mountain ranges. This was the first of five glaciations which followed one another in the course of the era, alternating with interglacial periods, during which the ice withdrew and temperatures rose above today's.

This fluctuation between cold and warm periods caused considerable variations in the fauna and flora. Large migrations from continent to continent took place, until the modern situation was reached.

The Quaternary is the era which gradually leads to the modern world, linking the past with the present. It is the era of man, as it was then that mankind established itself and organized itself culturally and socially. The Quaternary ended some ten to twelve thousand years ago, when the ice receded for the last time, marking the end of a long glaciation. It was then that the modern era began, the one in which we live. It is characterized, organically speaking, by the immeasurable expansion of mankind which has become responsible for the future of the Earth.

Specimens of Arctica islandica *from the Quaternary sediments near Palermo (Sicily). This large lamellibranch is a typical "cold guest". It came to the Mediterranean, via the straits of Gibraltar, together with the glaciations*

The beginning of the Quaternary era coincided with the first general cooling down of the climate which affected the whole of the planet's surface and brought about several alterations in what had been the environment during the last period of the Cenozoic.

During the Quaternary, the continents and the seas occupied the same positions we know today; no substantial geographical and geological changes took place over the past two million years, except perhaps for the lowering and rising of the sea level connected with the glaciations. The latter constitute the main characteristic of the Quaternary. The first general cooling down was only the beginning in a sequence of extremely cold periods, called glaciations, and warmer ones, called interglacial periods, during which the climate was decidedly milder than today's.

The complete skeleton of an Anancus arvernensis, *from the Lower Quaternary of the Arno valley (Italy). During certain periods of the Quaternary, many Italian regions were dotted with lakes. The sediments deposited in these lakes contain the remains of the large mammals living on the shores*

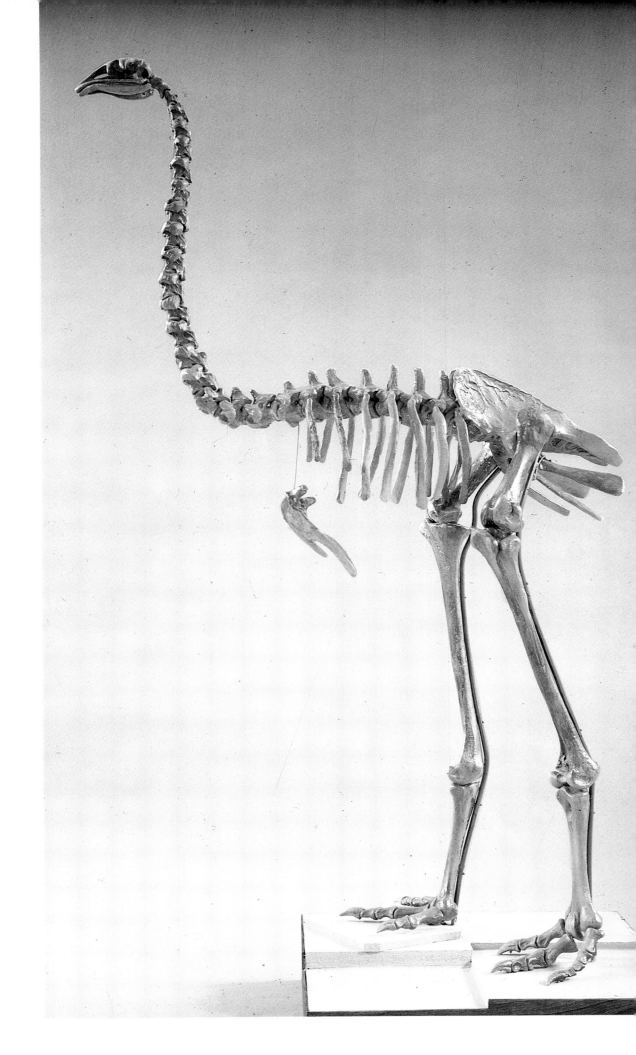

The skeleton of a Moa or Dinornis maximus, *from the Quaternary of New Zealand. It is one of the largest cursorial birds to be known, analogous to the modern ostrich. The Moas are believed to have survived until historical times. They are mentioned in Maori tales, and some specimens are so recent that they still retain part of their mummified skin and even feathers*

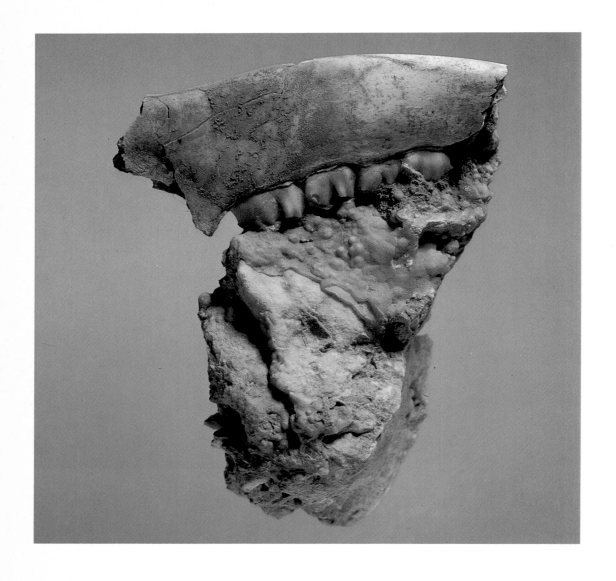

Cave sediments often contain fossilized organisms dating from the Quaternary. These sediments, sometimes called ossiferous breccias because of the large amount of specimens they contain, have yielded the remains both of animals living in the caves, such as bears, hyaenas and lions, and of those carried into them as prey. The specimen in the photograph comes from Zandobbio, near Bergamo (Italy)

During the glaciations, huge ice-caps formed on the poles and glaciers spread on the mountain ranges often joining in a continuous icy blanket. The margin of the northern ice cap covered part of Germany, all the Scandinavian countries, part of Great Britain and all of Canada, and reached further south than Chicago and New York. The Alps were completely covered by the ice blanket which, on the northern side, reached as far as the French plains, and in the south touched the Po valley.
The formation of such ice-caps obviously required huge quantities of water, which meant that the sea-level dropped and large areas of land emerged. To take one example, during the last glaciation the Italian peninsula was considerably modified: Sardinia and Corsica were joined together, Elba and some islands around Sicily were part of the mainland and the north part of the Adriatic was completely dry. At the same time,

A fossilized leaf in the travertine of Central Italy. Travertine is a rock which formed within springs and lakes from calcareous sedimentation. As it formed, it encased various organisms, particularly plants

the ice-caps eroded the surface forming, for instance, the glacial alpine valleys, the Italian lakes and the lakes of North America. The interglacial periods were warmer than the present climate and the ice melted. Ice-caps and glaciers withdrew, exposing the eroded valleys. Water returned to the seas, the latter's level rose and again covered part of the dry lands.
The alternation between glacial and interglacial periods is well documented, both by the layers of morainal sediments of varying dates and by the terraces formed on the coast-lines by the variations in sea-level. Geologists have proved that at least five main glaciations took place during the Quaternary: they are called, after European rivers: the Donau, Günz, Mindel, Riss and Würm. The last of them, ended between ten and twelve thousand years ago, marking the end of the Pleistocene and the beginning of the Holocene — the period in which we live.
The alternation between cold and warm periods had deep repercussions

on the fauna and flora, since the whole environment changed rapidly with the climate. During the cold periods, those European, North American and Asiatic regions which were not under ice supported northern-type forests and tundra, the temperate regions moved towards the equator and what is now desert must have been covered in luxuriant vegetation. Then came the interglacial periods, when what are now temperate areas were covered by subtropical vegetation. All this led to warm-type faunas succeeding cold-type ones throughout the Quaternary, as is clearly shown by fossils.

These successive faunas were not limited to the continents, as the seas were also deeply affected by the climate. Mediterranean marine molluscs are a typical example of alternating fauna: during the glacial periods the Mediterranean cooled down and warm-type fauna disappeared, while cold-type molluscs moved in via the Straits of Gibraltar from the Atlantic coasts. Then the interglacial period came and these cold molluscs were replaced by warm-type ones coming from the coasts of Senegal and Mauretania.

Palaeontologists call these molluscs "warm guests" and "cold guests". The discovery of these typical forms in a sediment allows the latter to be dated to a glacial or interglacial period. Examples of "warm guests" are *Strombus bubonius*, *Patella ferruginea*, *Cassis suburon* and *Conus testudinarius*. Important "cold guests" are *Arctica islandica*, *Mya truncata* and *Chrysodomus sinistrorsus*. On the basis of these alternating "warm guests" and "cold guests", geologists have divided marine sediments into glacial and interglacial ones and have correlated them with the sediments which had been deposited on the dry land during the same periods. The biological events of the Quaternary have thus been reconstructed almost completely, from the successions of the various environments to the main evolutionary developments.

The continental fauna was not very dissimilar from that of today, differentiations being found at specific levels and in distribution. In Europe, for instance, during the warm interglacial periods, the fauna consisted of elements which today characterize the intertropical areas. Great Britain was populated by elephants, hippopotamuses, felines and bovids, different species of which can today be found in the African savannahs. During the glacial periods, elks, gigantic deer and bears appeared, and the mammals of warmer climates moved to more southern

The cave bear is probably the best known of Quaternary animals. It was then extremely widespread in Europe, and its remains have been found in many localities. The specimen in the photograph was found in a cave in Lombardy (Italy)

187

regions. However, certain elements lived during the glacial periods which can no longer be found in cold areas: the mammoth, the woolly rhinoceros and the cave lion.

Parts of the planet were inhabited by characteristic faunas. South America, for instance, was characterized by a fauna of gigantic edentates, such as *Megatherium* (the giant ground-sloth), and the glyptodont, relatives of the much smaller, modern, South American edentates like the sloth and the armadillo. Huge land birds lived in New Zealand and Madagascar, such as the moa and *Aepyornis*, which laid the largest eggs ever found.

The large quantity of Quaternary fossilized remains can be explained both as a result of of the relative proximity of this era and because of the presence, in many parts of the globe, of sedimentary environments particularly favourable to the preservation of organisms. It is a well known fact that plant and animal remains stand a greater chance of preservation if they have spent a relatively short time within sediments: the shorter the time the greater the chance. It has also been ascertained that animal bodies and vegetable fragments need special conditions in order to fossilize. The fact that the vertebrates of the Quaternary are well known is due above all to the existence in many areas of the right environmental

Megaloceros, *the giant deer, was a typical inhabitant of the European plains during the Quaternary. It was a deer with gigantic horns, and was widespread across Europe and North Africa. The skull in the photograph was found in the river Po (Italy)*

The sediments of the river Po constitute a large deposit of interesting fossils. There are bones of vertebrates, rhinos, hippos, elephants, horses, deer and cattle dating from the Lower Pleistocene — from the Würm glaciation and the period immediately preceding it

conditions, such as lakes and swamps or caves, which prevent the organic remains from being immediately destroyed after death. The great numbers of fossilized Quaternary mammals found in central Italy are explained by the previous existence of large lakes, particularly in Tuscany and Latium. Local mammals had a chance of being preserved within such lakes, where the constant slow sedimentation of clay and sand protected them from destructive agents. One of the most interesting palaeontological documents of the era has been found in what were Tuscan lakes in the valley of the Arno: dozens of skeletons, thousands of fragments of elephants,

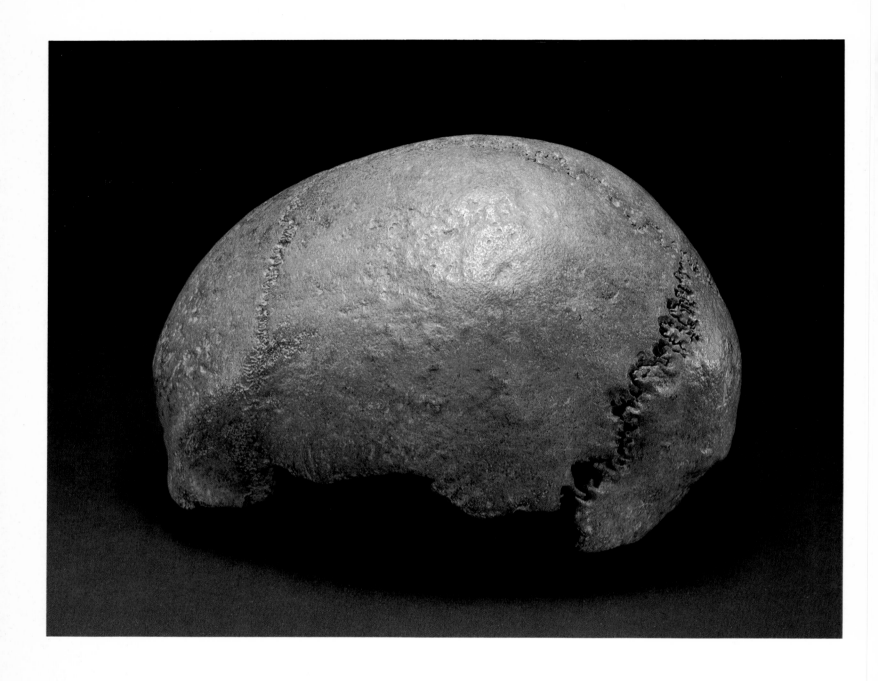

hippopotamuses, rhinoceroses, cattle and horses have allowed us faithfully to reconstruct life in the ancient forests and plains of thousands of years ago.

The last glacial period is that named after Würm, the end of which marked the beginning of the Holocene. As the climate became warmer, the fauna of the Pleistocene disappeared; bears, gigantic deer and woolly rhinos vanished leaving the last of the mammoth herds to linger on for a while in Siberia. In Europe, a few elements of the Pleistocene fauna survived the end of the glaciation, among them the aurochs, *Bos primigenius* which was still to be found around the year 1500.

We cannot end this chapter on the geological eras without mentioning that the Quaternary is also called the Anthropozoic era, since it was now that

Man is also present in the Po sediments. Some ten specimens have recently been found which can be ascribed to fossilized men with very primitive characteristics. They represent an important testimony to the presence of man in the Po valley during the Upper Pleistocene

Tools wrought by man appear for the first time in Quaternary sediments, and their evolution can be followed up to the present. Above: a smooth axe from New Guinea. Below left: an amygdaloid from the Lower Palaeolithic of Mauretania. Below right: the head of an Aztec javelin carved in obsidian

Ostriches and other savannah animals carved on a rock in the Sahara near Beni Abbes, Algeria. They indicate ancient climatic variations

human traces and human influence on the environment became increasingly meaningful.

The group of hominids from which modern man is supposed to descend had already appeared during the Cenozoic, over four million years ago. At the end of the Pliocene the australopithecines were widespread in Africa and left considerable traces. Palaeontological evidence shows that already towards the middle of the Quaternary mankind was well on the way to become the ruler of the planet. Fossil remains testify to the continued evolution of human types. Traces of human activity and tools become more numerous. In caves and on rocks cave paintings and other artistic manifestations appear more and more frequently.

At the end of the Pleistocene, mankind was widespread enough and well organized enough to be able, for the first time, to exert an influence on the environment. At the beginning of the Holocene it starts to exploit nature by means of agriculture and domestication.

Thus ends the prehistoric world, with the development of mankind. Now, other events begin, those told by History.

The Classification of Fossils

Systematics is the science concerned with the classification of organisms. It examines their characteristics and groups animals and plants within categories called systematic units. Such research aims at organizing the animal and vegetable worlds along certain lines, thus providing a framework of knowledge of the organisms. Upon such knowledge more complex studies are based, such as those concerned with evolution, ecology and biogeography. It would be quite impossible to study the evolution of a group of animals without prior identification of the organisms which belong to it, without having established its sub-groups (if any) or having identified the characteristics according to which, given organisms are classified within a given group.

Since fossils are the remains of plants and animals which were once alive, the systematist has to treat them in the same way as living organisms and classify them on the basis of the information supplied to him by the living world around him. He has to take into account individual variations as well as variations within groups, and all those organic characteristics which cannot be easily identified in the fossils but appear quite clearly in living specimens.

The fundamental systematic unit is the species, the only natural category. It can be defined as a group of organisms which can, potentially or effectively, interbreed. This definition has obviously little to do with palaeontology as it cannot be verified with the kind of material a palaeontologist has at his disposal. A palaeontological species is therefore more difficult to define, both because it cannot be verified and because the time factor (of little relevance in the case of living organisms) is extremely important to a palaeontologist.

A biologist's vision of the living world is limited to a single short period, while the palaeontologist has to consider a wider picture and look at each group of animals or plants as they evolve through the various stages of their history. A biological species can be defined only spatially; a palaeontological one (also called *chronospecies*) can also be defined on a time basis. Furthermore, a biological species only varies in space, while a palaeontological one varies in space and in time. Any definition of a palaeontological species is therefore very difficult: it could be reached in a spatial sense, for each period, even though it would still remain a fairly subjective definition which cannot be checked; but a temporal division can only be entirely subjective. A palaeontologist is often faced with groups which constantly and gradually change in the course of time, and is therefore quite unable to establish any systematic divisions within a given evolutionary line.

Systematic units higher than the species are not based on characteristics which can be biologically proven, but on subjective interpretations or traditions. Species are thus grouped into *genera*, genera into *families*, the latter into *orders*, orders into *classes* and these into *phyla* (singular: *phylum*). Many other systematic categories exist besides these main ones: superfamilies, for instance, and subphyla, sub-orders, subspecies, and so on.

Systematics are governed by very strict rules. All

researchers must follow them in order to avoid the utter chaos which would result from each using different nomenclature. These rules are outlined in the *International Code of Zoological Nomenclature*, published in 1961 and followed by all scientists. This publication outlines the rules which must be followed in order to produce a correct nomenclature of the organisms, as well as their appropriate description.

Each organism, whether animal or vegetable, is indicated with a double Latin name, the first half of which indicates the genus and second the species. The generic name is spelled with an initial capital, while the specific one is always in lower case, even when derived from the name of a scientist. The scientific name of a lion, for instance, is *Panthera leo*; it is followed by the name of the author — the scientist who first defined and described the species — and by the year in which the description was first published. One can thus fairly easily refer to the first publication. In the case of the lion, we would thus have: *Panthera leo* (Linnaeus, 1758), where the parentheses before and after the name of the author and the year of publication mean that the author had actually assigned the species to a different genus from the present one. Linnaeus had classified the species *leo* as belonging to the genus *Felis*, not the genus *Panthera*, which was introduced later as a result of further systematical research.

As an example of the complete classification of an animal let us see what happens to our lion:

species	*leo*	class	*Mammalia*
genus	*Panthera*	subphylum	*Vertebrata*
family	*Felidae*	phylum	*Chordata*
order	*Carnivora*		

Which means that the lion is a carnivorous mammalian vertebrate belonging to the cat family. If we take the tiger and the leopard, the scientific names of which are respectively *Panthera tigris* (Linnaeus, 1758) and *Panthera pardus* (Linnaeus, 1758), we realize that lion, tiger and leopard all belong to the same genus. Should the case arise, during the course of classification, where the genus, but not the species, can be defined, the generic name will be followed by the letters "sp" or "ind.sp.", indicating an "indeterminate species". On the other hand, should the attribution of a species be uncertain, the specific name can be preceded by a question mark or by the letters "cf", for "compare", thus indicating that only a comparison has been made, without any attempt at a precise classification.

Let us go back to the lion and imagine that an inexperienced zoologist, having captured one of these animals, is unable to classify it. He would then indicate it as *Felis* sp., that is a feline with no specific name. However, should he write *Felis* cf *leo*, he will mean that he is not certain he has found a lion, even though the animal looks very much like one.

The following pages give a systematic list of the major groups of plants and animals with fossil representatives. The information has been kept at the level of the orders and only the main characteristics of each group have been given, together with the boundaries of their temporal distribution.

PLANTS

This list does not include fungi and lichens which are extremely rare as fossils.

Phylum **Schizophyta**

Class Schizophyceae

The Schizophyceae, or blue-green algae also called Cyanophyceae, are filament-like organisms with a tough sheath and the ability to fix the calcium carbonate in water thus forming calcareous encrustations. These concentric encrustations have been known to go back, as fossils, to the Archaeozoic era, in the ancient seas of which they formed extensive calcareous banks. These fossil structures are called stromatolites (figure 1).

Fig. 1

Class Schizomycetes

This class includes those microorganisms commonly known as bacteria. They are very simple organisms, colourless and usually lacking chlorophyll. Bacteriological existence in past geological epochs has been indirectly demonstrated by the fact that some sedimentary deposits of iron, limestone and phosphates have been indirectly attributed to their activity. Fossilized bacteria have been identified in coal deposits, in some extremely ancient rocks, in excrements and soft teguments of certain fossilized animals. The oldest known organism, *Eobacterium*, is assigned to this class; it was found in the rocks of Fig Tree, in eastern Transvaal, which are over three billion years old.

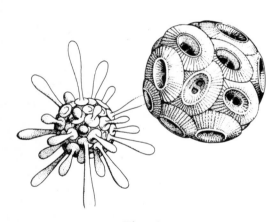

Fig. 2

Phylum **Chrysophyta**

Class Chrysophyceae

Two groups of microorganisms, very common in the fossilized state, belong to this class: the Coccolithophoridae and the Silicoflagellata. The Coccolithophoridae (figure 2), at present widespread in warm waters, are microscopic flagellate cells covered in tiny, variously shaped, calcareous plates. The plates are preserved as fossils, as they deposited on the sea-bed after the death of the cell and contributed to limestone deposits. Coccolithophoridae have been known since the Palaeozoic: during the Cretaceous they were so numerous that the chalk of the Paris basin and of South-east England can be regarded as being made almost entirely of the remains of such organisms. The silicoflagellates were not so numerous (figure 3). They were planktonic microorganisms typical of cold seas, consisting of a siliceous skeleton made of two reticulated valves. They first appeared during the upper Cretaceous and became widespread mainly during the Cenozoic.

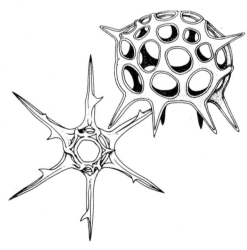

Fig. 3

Phylum **Pyrrophyta**

Class Dinophyceae

The class Dinophyceae includes the dinoflagellates (figure 4), planktonic organisms living at present in fresh waters and marine waters; they are encased in a tough theca dotted with spines and variously shaped excrescences. Fossils of the theca have been found, dating back to the Palaeozoic.

Phylum **Bacillariophyta**

Class Bacillariophyceae

The Bacillariophyceae include microorganisms, protected by a siliceous shell, which live in both fresh and marine waters of cold areas and are known as siliceous algae or diatoms (figure 5). Their siliceous shells, extremely variable in both shape and size, have contributed to the formation of sedimentary deposits in lagoon and lacustrine environments. These white and friable rocks consist almost exclusively of these shells and are called fossil flour, or diatomite. The oldest known diatoms have been found in sediments dating from the Lower Jurassic.

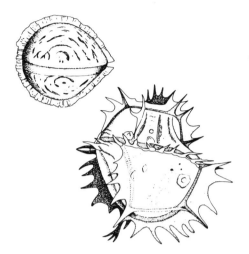

Fig. 4

Phylum **Phaeophyta**

Class Phaeophyceae

The Phaeophyceae are the brown algae at present widespread both in the seas and in many continental fresh waters. They are not commonly found as fossils. Some large fragments attributed to them have been found in Silurian and Devonian rocks. They are fragments of stem, almost a metre in diameter, which were at first thought to be the trunks of gigantic conifers.

Phylum **Rhodophyta**

Class Rhodophyceae

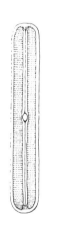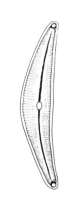

Fig. 5

These are the red algae at present widespread in all seas. the Solenoporaceae, living from the Silurian to the Cretaceous, were rather important in their day, as were the Corallinaceae, which first appeared during the Cretaceous and are still present. Seen through the microscope, they appear as radial rows of small cells which, as a conglomerate, form nodular masses or encrustations. The latter contributed to the formation of underwater reefs, either in isolation or in association with coralliferous deposits.

Phylum **Chlorophyta**

Class Chlorophyceae

The Chlorophyceae are the green algae which today live in tropical seas and fresh waters. In the past, they played a very important part in the formation of calcareous cliffs, either in isolation or together with other similar organisms, such as the coelenterates. From the palaeontological point of view the most interesting groups are the Codiaceae

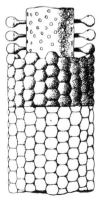

Fig. 6

Fig. 7

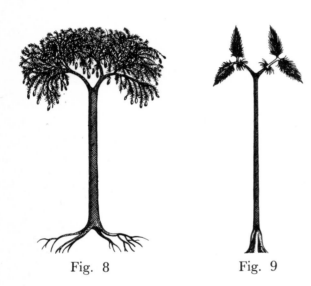

Fig. 8   Fig. 9

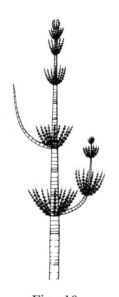

Fig. 10

and the Dasycladiaceae, which both appeared during the Silurian. The Codiaceae are formed by branched tubes subdivided into tiny articles one above the other, each covered in small holes. the Dasycladiaceae (figure 6) are microscopic algae formed by a branched thallus which, on the outside, has the property of fixing calcium carbonate dissolved in the surrounding water thus producing a kind of external calcareous skeleton.

The large amounts of these algae have made a considerable contribution to the formation of Triassic coral reefs in northern Italy.

Phylum **Pteridophyta**

The Pteridophyta are land-plants without flowers or seeds which reproduce themselves through spores. First appearing during the Silurian, they constituted the basic terrestrial vegetation throughout the Palaeozoic. Their earliest representatives, the Psilophytales, were the first to conquer the dry land.

Class Psylopsida

Together with some types still alive today, this class includes the group of the Psilophytales, the first land plants. They first appeared during the Silurian and became widespread during the Devonian. They were small plants, consisting basically of a subterranean thallus without roots from which dichotomously subdivided vertical aerial branches orginated (figure 7).

Class Lycopsida

This large class is today represented by a few herbaceous forms, the lycopods, but some of the most striking extinct plants belonged to it: during the Carboniferous, the period of their maximum development, they could reach a height of 30 m. They were the Lepidodendrales and the Sigillariaceae which, during the Palaeozoic (from the Devonian to the Permian) formed thick forests by swamps and lakes. The Lepidodendrales (figure 8) were formed by a trunk with many branches at the top. The Sigillariaceae (figure 9), on the other hand, had few branches. The whole of the trunk was covered in narrow, elongated leaves which left a scar as they fell. The shape of these scars, always well preserved as fossils, is the basis of the classification of these plants.

Class Sphenopsida

This class is today represented by Equisetales the horse-tails (small leafless plants). To it are attributed certain tree-like forms of the Palaeozoic which could attain considerable height (figure 10). These were the Calamariaceae, plants living with their basal part immersed in the water of the swamps, which contributed to the formation of the large marshy forests of the Carboniferous. These ancient plants had an articulated trunk, each segment being joined by nodes which were easily separated. The trunk ended in a tuft of pseudo-leaves, while other small leaves sprang from the joints of the nodes.

Class Pteropsida

These are the ferns, still abundant today, which were widely distributed in the Palaeozoic both as small herbaceous plants and as tree-like forms reaching considerable heights (figure 11). They were the first plants to develop real leaves, which allowed them to be more independent of water and to form an important part of the large Carboniferous forests growing in drier climates.

Phylum **Spermatophyta**

The spermatophytes are the most highly evolved of plants since, unlike the previous ones, they reproduce themselves by means of seeds.

Subphylum Gymnospermae

The Gymnospermae do not use spores for reproduction and their seeds are enclosed in a protective case. They first appeared during the Devonian.

Class Pteridospermae

The Pteridospermae (figure 12) are now completely extinct. They were very similar to ferns but distinguished by the presence of proper seeds. They first appeared in the Devonian and constituted a large part of the Carboniferous vegetation. Their importance grew increasingly after this period, and during the Permian they represented the main characteristic of the two large vegetative areas: the flora of the northern continents formed by pteridosperms, cycads and cordaites, and the Gondwana flora consisting of pteridosperms of the genus *Glossopteris*. The latter was widespread and characterized by oval leaves. The Pteridospermae disappeared at the end of the Palaeozoic.

Class Cycadophytae

These plants first appeared during the Carboniferous and are now represented by about one hundred tropical species very similar to palms. The stem is usually unbranched, cylindrical and covered in scales. A tuft of leaves similar to that of palms grows from the crown (figure 13). The Cycadophytae flourished mainly during the Triassic and Jurassic.

Class Bennettitales

First appeared during the Permian. These plants reached their maximum development in the Jurassic and Cretaceous and became extinct at the beginning of the Cenozoic. They were similar to the Cycadophytae, with a cylindrical, unbranched trunk and a tuft of pinnate leaves on the crown, which could be as much as 3 m long. They were however distinguished by their flower which carried both the male and female organs. Many scientists regard them as the group from the which the Angiospermae derived (figure 14).

Class Cordaitales

These primitive Gymnospermae first appeared during the Devonian and became extinct at the end of the Palaeozoic. They could grow up to 30 or 40 m. Their trunk was slender and their branches covered in leaves. The latter were elongated and set spirally on the branches. They had parallel veins and could be up to one metre long (figure 15). They probably looked very similar to conifers.

Fig. 11

Fig. 12

Fig. 13

Fig. 14

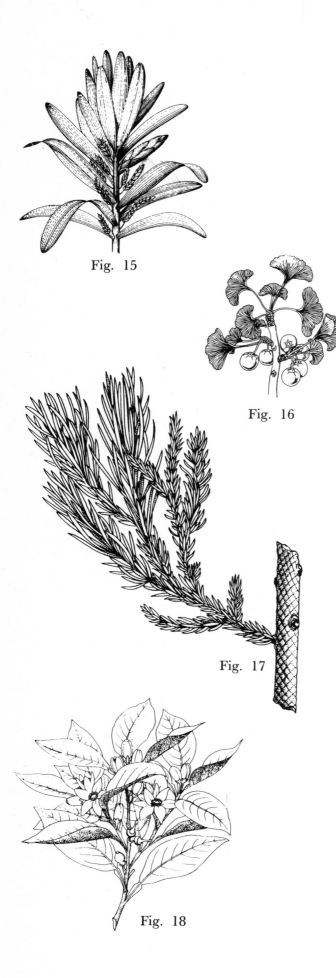

Fig. 15

Fig. 16

Fig. 17

Fig. 18

Class Ginkgoales

The Ginkgoales, only one species of which survives today (*Ginkgo biloba*, figure 16), first appeared during the Carboniferous and reached their maximum development in the Jurassic. They have a tall stem and leaves easily recognizable by their lobate shape.

Class Coniferae

Conifers are well known and widespread all over the world. They first appeared during the Carboniferous and established themselves during the Permian and Triassic. The first conifers, among them the genus *Walchia* (figure 17), were very similar to the modern araucaria (monkey-puzzle or Chile pine).

Subphylum **Angiospermae**

These are the higher plants, having seeds contained within an involucre as well as flowers. Today they form the majority of terrestrial vegetation. Their origin is unknown: it is thought they may go back to the Triassic but their diversification only began during the Cretaceous.

Class Dicotyledoneae

The embryo of these plants is provided with two cotyledons. They first appeared during the Mesozoic and became predominant during the Tertiary; today they form the largest part of our flora (figure 18).

Class Monocotyledoneae

These are the most advanced of plants and their seeds only have one cotyledon. They include the oldest known angiosperm, *Sanmiguelia lewisi* (figure 19), which has been found in the Triassic deposits of Colorado. The monocotyledons, among which are the palms and the grasses and sedges, appear to have evolved rapidly at the end of the Mesozoic. Several fossil palms have been found in the Italian Tertiary deposits.

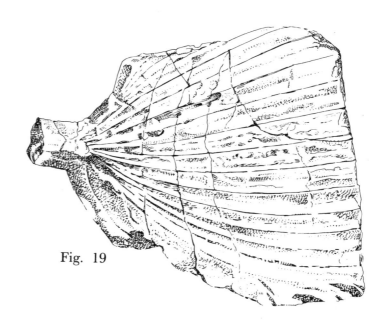

Fig. 19

ANIMALS

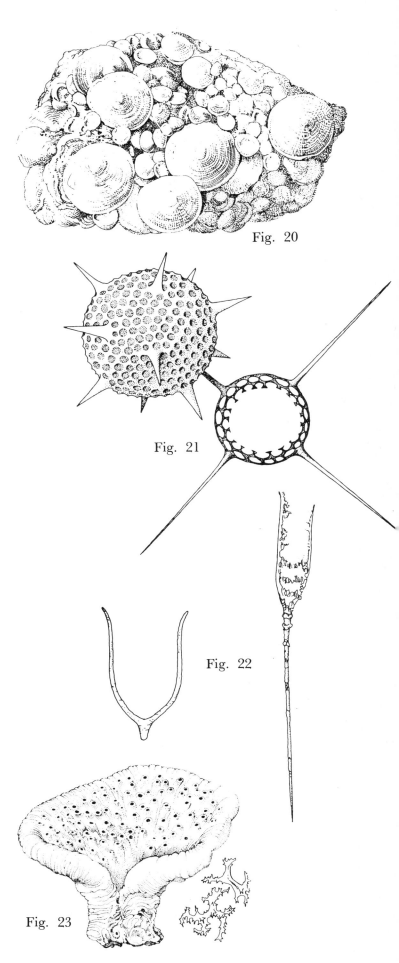

Fig. 20

Fig. 21

Fig. 22

Fig. 23

Phylum **Protozoa**

The Protozoa are of variable dimensions, from one micron to several centimetres. They are mainly aquatic animals formed by a single cell. Only those endowed with tough structures, such as a calcareous, chitinous or siliceous shell, are preserved as fossils. They are divided into numerous classes and an exceptional number of species.

Class Sarcodina

Order Foraminifera

Unicellular organisms, mainly marine ones, with a calcareous or chitinous shell, the foraminiferans have supplied us with a large number of guide fossils. We should mention those known as macroforaminiferans, which sometimes grew to considerable dimensions and had complicately structured shells. See, for instance, the nummulites (figure 20) of the Palaeogene, the fusulinids of the Carboniferous and Permian. Foraminiferans first appeared in the Cambrian and are still alive today.

Order Radiolaria

These are marine unicellular organisms with a perforated siliceous shell, varying in shape and studded with needles and spines (figure 21). The Radiolaria appeared in the Cambrian and the layers of their fossilized skeletons have formed huge rocky conglomerates.

Class Ciliata

These are aquatic protozoans with a shell often preserved as a fossil. The Ciliata most frequently found in this state are the Tintinnidae, with their cup- or bell-shaped shells (figure 22); they first appeared during the Jurassic.

Phylum **Porifera**

The Porifera, or sponges, have been preserved as fossils since the Cambrian. They have been found either complete or as siliceous or calcareous fragments representing the microscopic spicules which formed the internal supports of their bodies (figure 23).

Phylum Archaeocyatha

These organisms are now extinct, having lived in the seas of the Cambrian. They were conical animals, consisting of two porous and concentric walls connected by vertical septa and horizontal planes (figure 24). The central part of the organism was a cavity, similar to the pseudogastric cavity of the sponges. They lived implanted on the sea-bed in shallow waters and near the coast. They have been variously classified as coelenterates (owing to their vertical radial septa) and as sponges (due to their central cavity). The are today considered to be a separate phylum. They are represented by some of the oldest Italian fossils found in the Cambrian sediments of Sardinia.

Phylum **Coelenterata**

The coelenterates were aquatic animals consisting, basically, of a central cavity with a single opening, surrounded by a ring of tentacles. They can be found in two forms: a sessile one, or polyp; and a mobile one, or medusa. Both forms can alternate in the life of a single species. Coelenterates have been known, as fossils, ever since the Archaeozoic. Sediments of that era have yielded the traces left by the mantles of medusae. The polyp forms appeared later, during the Ordovician. Coelenterates provided with a calcareous skeleton formed large colonies during the latter period and began to build coral reefs which, transformed and displaced by later geological phenomena, have given origin to mountain ranges. An example of this are the Dolomites — the remains of coral reefs which originated during the Triassic.

Class Protomedusae

These are primitive, now extinct, medusae, the traces of which have been found in many Cambrian deposits as well as in pre-Cambrian rocks of North America. They are the oldest known coelenterates, similar to modern medusae but without tentacles.

Class Dipleurozoa

The Dipleurozoa are medusae which lived during the Cambrian and are now extinct. They consisted of a bell-shaped mantle with a central opening and radial lines, and edged with several filiform tentacles (figure 25).

Class Scyphozoa

The Scyphozoa are coelenterates still alive today, which can seldom be found as fossils. They lack hard parts and are known mainly from the prints left in the sediments by the mantle of the medusae. The oldest among them date from the Cambrian. A group of strange organisms, now extinct, which lived from the Cambrian to the Triassic, are classified as Scyphozoa. They are the Conularia, with a pyramidal external skeleton, between 4 and 10 cm long, within which the animal lived (figure 26).

Class Hydrozoa

These coelenterates are still living today and can be found either as polyps, alone or in colonies, or as medusae. The polyps often present a more or less calcified skeleton and have therefore been well preserved as fossils. Various groups of Hydrozoa played an important role in the construction of reefs: among them the millepores of the Cretaceous, the stromatopores of the Cambrian to Cretaceous, and the Spongomorphida of the Triassic and Jurassic.

Class Anthozoa

The Anthozoa are a group of coelenterates which originated in the Ordovician and specimens of which are now widespread in all the seas. They include forms without a skeleton (the sea anemones) and others with a calcareous skeleton, some of which live in isolation and

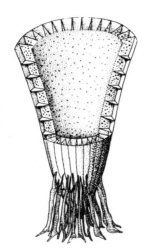

Fig. 24

Fig. 25

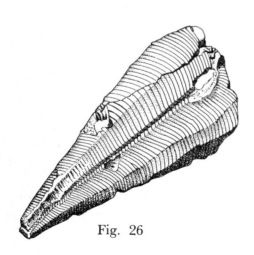

Fig. 26

others in colonies (figure 27). The Anthozoa with a calcareous skeleton, among which are red corals and madrepores, formed huge coral cliffs now transformed into rocky conglomerates which still hold the fossilized remains of their "builders". The Anthozoa are divided into various groups: the Octocorallia, with a moderately strong skeleton (among them the red corals), the Zoantharia, representing the majority of both fossilized and living corals, and the Tabulata now extinct but once living in the seas of the Palaeozoic.

Phylum **Bryozoa**

The Bryozoa live in colonies and resemble, in their external form, the coelenterates; they differ from the latter in their much more complicated anatomy. They are characterized by a complete alimentary duct, with mouth, stomach and anal opening. Their colonies exude a thin calcareous skeleton and build branched structures of various elegant shapes. Each colony consists of many individuals, each one of which lives within a small cell in the skeleton (figure 28). Bryozoa are common in modern waters and were even more widespread in the past. They first appeared during the Ordovician and, in certain periods and certain seas, they became so numerous that the accumulation of their skeletons formed sandy layers which, once solidified, became large rocky formations.

Phylum Brachiopoda

The brachiopods are benthic animals with a bivalve shell and complex anatomy which brings them close to the Bryozoa, notwithstanding the considerable superficial and morphological differences (figure 29). The shell enabled them to be preserved easily as fossils. The brachiopods appeared during the Cambrian, became widespread during the Palaeozoic and Mesozoic, and fell into a decline during the Cenozoic. Only a limited number of species are still living today. The phylum is divided into two classes: the most primitive are the inarticulated brachiopods, with a chitinous or calcareous shell (the two valves, lacking a hinge, are held together by two muscles); the second class, the articulated brachiopods, have a calcareous shell with a hinge in the form of teeth and notches.

"Worms"

The animals included under this very general heading actually belong to different phyla. They are organisms with a soft body, rarely preserved as fossils and therefore mainly known from the traces and tracks they have left within the sediments as they moved about. From a palaeontological point of view, the most interesting groups are the phyla Chaetognatha and Annelida and the conodonts.

Phylum Chaetognatha (the arrow-worms)

Only one fossilized representative of these marine animals is known, the genus *Amiskwia*, found in the Cambrian deposits of British Columbia (Canada).

Phylum **Annelida**

These segmented worms are represented, in the fossilized state, by members of the class Polychaeta, comprised of three orders.

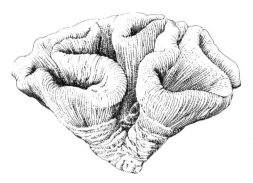

Fig. 27

Fig. 28

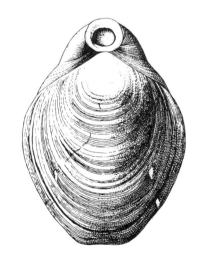

Fig. 29

Order Errantia

The Errantia are not well known as fossils. They had a soft body and are known, from the Ordovician onwards, mainly by the remains of their jaws, called scolecodonts.

Order Sedentaria

These annelids have left various traces of their existence in the seas of the past: dwelling tubes, holes bored in the sea-bed and calcareous traces of dwellings (figure 30).

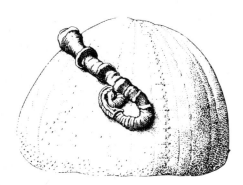

Fig. 30

Order Myskoidea

These large marine annelids lived during the Cambrian and became extinct at the end of the period. They had a soft body and specimens have been found preserved in the Burgess Shale (Canada).

Conodonts

They are one of the last major palaeontological mysteries. They are microfossils similar to tiny jaw plates and are very common in rocks of marine origin which formed in the interval between the Ordovician and the Triassic (figure 31). An intact fossil of a "conodont-animal" was recently discovered in the Carboniferous of Scotland. The animal was a worm-like form and the conodonts were its jaws. However, the systematic position of this animal is still unclear.

Phylum **Arthropoda**

The arthropods are the most common invertebrates today and among the most specialized ones. Their external, chitinous and resilient skeletons have allowed them to fossilize quite well. They are known to have existed since before the Cambrian.

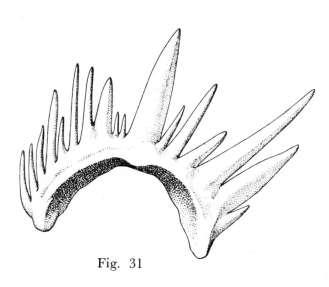

Fig. 31

Subphylum Onychophora

Known to have existed since the Cambrian, the Onychophora (figure 32) first lived in a completely different habitat from their present one. One of the earliest known onychophorans, the genus *Aysheaia* of the Cambrian, was a marine animal. Today, these animals live in the damp undergrowth of tropical rain-forests.

Subphylum **Trilobitomorphida**

Class Trilobitoida

The trilobitoids are primitive arthropods which lived during the Cambrian and which, on the basis of certain characteristics, can be considered the ancestors of various groups of arthropods.

Class Trilobita

The trilobites, Palaeozoic arthropods which appeared in the Cambrian and became extinct in the Permian, were pelagic and benthic animals characterized by an exoskeleton subdivided into three parts or lobes, both longitudinally and transversely (figure 33). Each segment was provided with a pair of primitive legs, all alike except for the first pair which had developed into antennae. On the basis of perfectly fossilized specimens, the anatomy of trilobites is known in great detail. The exceptional amount of fossil remains has rendered these animals extremely useful in dating Palaeozoic deposits and has enabled scientists to reconstruct their evolutionary history.

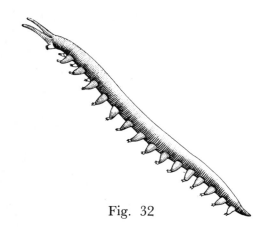

Fig. 32

Subphylum **Chelicerata**

Class Merostomata

These aquatic animals are today represented by the genus *Limulus* (the king-crab) which first appeared during the Triassic. In the Palaeozoic the Merostomata were very common, particularly the group Eurypterida — huge water scorpions which in some cases grew to enormous proportions (figure 34). The genus *Pterygotus*, which lived in the Silurian and Devonian, was about 2 m long and the largest arthropod which ever existed.

Class Arachnida

These were among the earliest land animals. Terrestrial scorpions appear in the late Silurian and several types of spider and mite are known to have been present by the Middle Devonian.

Class Pycnogonida

Very scarce as fossils. The first representatives appeared during the Devonian.

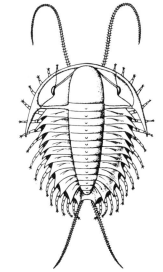

Fig. 33

Subphylum **Antennata**

Class Crustacea

The crustaceans first appeared during the Cambrian and are quite common as fossils. Among those which became most widespread are: the ostracods, small crustaceans with a chitinous or calcareous shell consisting of two valves, which are used as zone-fossils in the dating of deposits; the cirripedes (barnacles), with a body protected by a thick shell of calcareous plates, which lived anchored to coastal rocks; the malacostracans, the most evolved of crustaceans, which included prawns and crabs, and which first appeared respectively in the Triassic and Jurassic (figure 35).

Class Myriapoda

The myriapods appear as millipedes in the Silurian and as centipedes in the Devonian, and are abundant by the Carboniferous. A remarkable early myriapod was the Carboniferous form *Arthropleura* which grew to 2 m in length.

Class Insecta

The earliest known insects are the small wingless springtails from the Lower Devonian of Scotland. By the Upper Carboniferous there were many orders of winged insects including the Orthoptera (bushcrickets), Blattodea (cockroaches), Palaeodictyoptera (early sap-sucking insects) and Protodonata (early dragonflies including *Meganeura* with a 70 cm wing span). A wide range of small insects have been preserved in Cretaceous and Tertiary ambers.

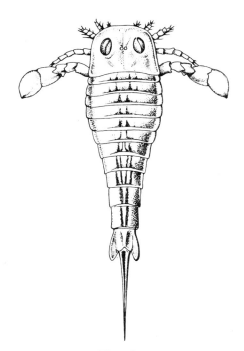

Fig. 34

Phylum **Mollusca**

No group of invertebrates, however large, will ever have the palaeontological importance that molluscs have, both for their abundance and diversity and for the information which their fossils supply scientists with. Their abundance as fossils is due to the fact that most of them have a calcareous shell which is very easily preserved. They are particularly interesting from a palaeontological point of view

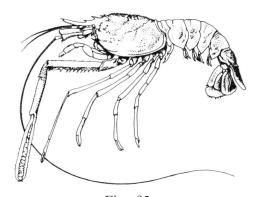

Fig. 35

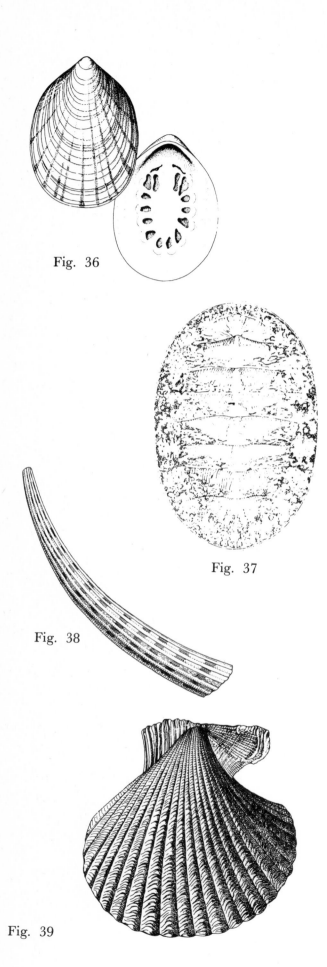

Fig. 36

Fig. 37

Fig. 38

Fig. 39

because of the large numbers found within sediments and because they represent many different environments. The numerous different forms into which they evolved in the course of time make them excellent guide fossils in the relative dating of sediments. Since they enable us to reconstruct past environments, they are also excellent ecological gauges.

Class Monoplacophora

These animals are characterized by an external shell consisting of a single valve in the shape of a hood. They are the most primitive of all molluscs (figure 36). They were common during the Palaeozoic and were believed to be extinct until a living specimen was found in 1952 while dredging the Pacific sea-bed.

Class Polyplacophora

The polyplacophorans are marine molluscs the shell of which consists of eight articulated plates (figure 37). They are not very common as fossils and the oldest among them have been found in Cambrian deposits.

Class Scaphopoda

These molluscs have a calcareous external shell in the shape of a slightly curved tube or tusk (figure 38). They first appeared during the Ordovician and are still quite common in our seas.

Class Lamellibranchia

The group of the Lamellibranchia (figure 39), to which oysters, mussels and other edible molluscs belong, is extremely interesting from a palaeontological point of view because specimens are easily found as fossils and their presence within the sediments is the source of extremely useful ecological information. The lamellibranchia first appeared during the Cambrian and, in the course of evolution, continued to give rise to new groups, some of which became extinct after a more or less long life. Others have survived almost unchanged to the present day. The extinct groups included the rudists, which lived on reefs and had a shell consisting of two different valves. They lived during the Jurassic and Cretaceous.

Class Gastropoda

Gastropods (figure 40) are also very important to palaeontologists. They evolved in the Cambrian and gave rise to a large number of diverse species which palaeontologists use as zone-fossils and ecological and climatic indicators. Among those groups which are now extinct we should mention the tentaculites (fam. Tentaculitidae), small gastropods which lived during the Lower Palaeozoic and were characterized by a conical shell.

Class Cephalopoda

Undoubtedly the most evolved of molluscs. They are exclusively pelagic animals, almost all of them with an external or internal shell. They have been very common in all the geological eras. Of the three subclasses into which they are divided, only the Dibranchia are widespread today, while the Nautiloidea are only represented by a few species of the genus *Nautilus* and the Ammonoidea disappeared at the end of the Cretaceous.

Subclass Nautiloidea

The nautiloids (figure 41) are represented in modern seas by the genus *Nautilus*, characterized by a spiral chambered shell. They were very common in the past. During the Palaeozoic and in the Cambrian, species could be found with a straight shell, a slightly curving shell and a spiral shell. They began to decline during the Mesozoic, when the most primitive types vanished and the modern ones, analogous to our *Nautilus*, first appeared.

Subclass Ammonoidea

The ammonoids (figure 42), usually called ammonites, are a large group of cephalopods which appeared during the Devonian and became extinct at the end of the Cretaceous. They were pelagic animals characterized by an external shell, spiralling on a plane, divided internally into chambers by complex septa. The large numbers of different forms, their rapid evolution and wide distribution make the ammonites excellent guide fossils for the Mesozoic sediments.

Subclass Dibranchia

This subclass includes almost all the cephalopods living today: cuttlefish, squids and octopuses, together with the extinct group of the belemnites, or belemnoids. The latter were very numerous during the Mesozoic (having first appeared in the Triassic) and became extinct towards the end of the Cretaceous or, perhaps, at the beginning of the Palaeogene. These cephalopods looked like squids with ten tentacles and had a very strong internal shell, usually the only part of the animal to be preserved as a fossil (figure 43). As for the groups which are still living today, the cuttlefish appeared at the end of the Mesozoic, the squids during the Lower Jurassic and the octopuses during the Upper Cretaceous.

Phylum **Echinodermata**

The echinoderms are complex marine animals with well differentiated nervous, circulatory and digestive systems. They are all characterized by an external skeleton which can be continuous, consisting of calcareous plates, or discontinuous, consisting of isolated spicules. In the former, and by far the most frequent case, the skeleton encloses all the organs and, being very tough, is easily preserved by fossilization. The echinoderms appeared during the Cambrian and developed through very complex evolutionary stages. They are divided into several classes, some of which are now extinct.

Class Cystoidea

These echinoderms lived from the Ordovician to the Devonian and were characterized by a globular or flattened theca consisting of polygonal plates (figure 44). The theca also had a certain number of arms and a peduncle which anchored the animal to the solid substrate of the sea-bed.

Class Blastoidea

The blastoids lived from the Ordovician to the Permian and were

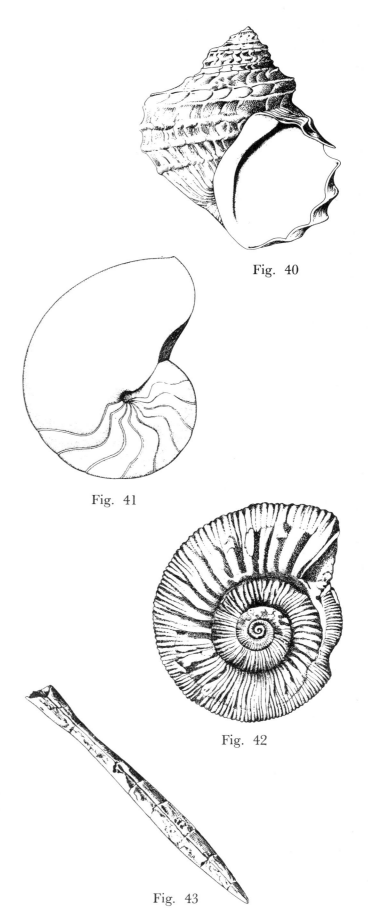

Fig. 40

Fig. 41

Fig. 42

Fig. 43

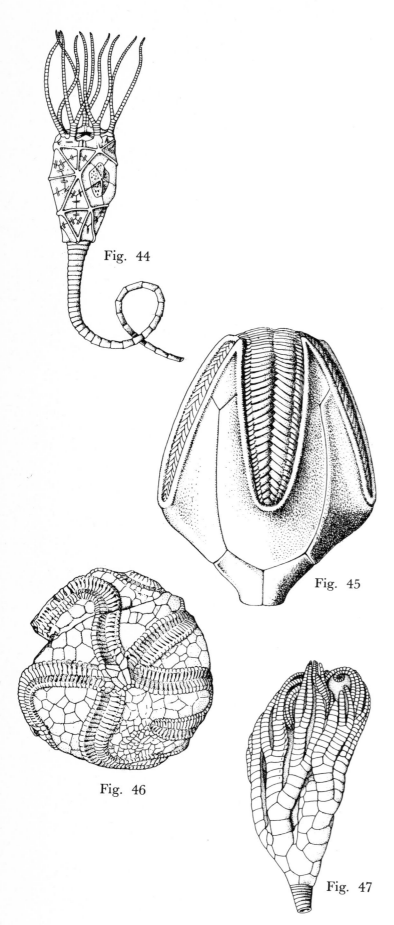

Fig. 44

Fig. 45

Fig. 46

Fig. 47

characterized by an ovoid theca consisting of thirteen main plates with neither arms nor peduncle (figure 45).

Class Edrioasteroidea

Their bodies were either globular or flattened and lived anchored to the sea-bed by the lower part of their theca, which had no peduncle. The theca itself consisted of a large number of irregularly shaped plates and of five sinuous locomotive areas formed by smaller plates (figure 46). The edrioasteroids lived from the Cambrian to the Carboniferous.

Class Crinoidea

These are the modern sea lilies, consisting of an ovoid theca with articulated arms and a long peduncle which anchored the animal to the sea-bed (figure 47). The crinoids appeared during the Ordovician, became extremely numerous during the Palaeozoic and have a very complex evolutionary history.

Class Stelleroidea

These are the common star-fish, the evolution of which began during the Ordovician. Many specimens have been found perfectly fossilized (figure 48).

Class Echinoidea

These are the well-known sea urchins, which were much more common in the past than at present. Their history began during the Ordovician. They are divided into two fundamentally different types: the regular echinids (figure 49), i.e., the common sea urchins, with a regularly structured theca, and the irregular echinids (figure 30), living in sand rather than on rocks, with an irregular theca and lacking large spines.

Class Holothurioidea

They are represented by the modern sea cucumber. These echinoderms lack a continuous external skeleton formed by plates, but possess, inside their bodies, several calcareous sclerites characteristically shaped and therefore easily recognizable (figure 50). The absence of a tough theca means that they are rarely found as complete fossils. Isolated sclerites are however quite common within sediments. The first individuals appeared during the Cambrian.

Subphylum Stomochordata

These are transitional animals between invertebrates and chordates.

Class Pterobranchia

They are organisms living in colonies, in which the various individuals are united by a common structure, a branching chord called a stolon. Both the stolon and the individuals forming the colony are protected by a skeleton in the shape of a ramified tube. They are known from living forms and from Cretaceous and Tertiary fossils.

Class Graptolita

The graptolites are extinct but were extremely numerous in the seas from the Cambrian to the Lower Carboniferous. They were pelagic organisms living in colonies, like the pterobranchs, and were protected by a chitinous exoskeleton. The colony consisted of small annular segments, superimposed, and extended along a straight or curved axis in either simple or branched form; it supported the thecae containing the various individuals (figure 51).

Phylum **Chordata**

Chordates are characterized by bilateral symmetry, a dorsal nervous system and a fibrous rod — the notochord — which extends along the back and serves as a primitive skeletal structure in the lower chordates. The chordates are a very diverse group ranging from the tunicates (sea-squirts) and cephalochordates (the lancelets), both small benthic marine forms which are scarce as fossils, to the vertebrates which possess a complex internal skeleton and are abundant in all environments.

Subphylum **Vertebrata**

Vertebrates are characterized by an internal skeleton composed of cartilage or bone including a segmented vertebral column surrounding the nerve-cord and incorporating or replacing the notochord of the lower chordates. The front end of the nerve-cord is expanded as a brain which is enclosed and protected by a cephalic skeleton or skull. Vertebrates are larger and more active than primitive chordates and possess a circulatory system with an anterior pump or heart to distribute oxygen and nutrients throughout the body more efficiently.

Class Agnatha

These are the most primitive grade of vertebrates which retain a cartilaginous internal skeleton and lack jaws. They first appear in the Upper Cambrian and are very abundant through the Silurian and early Devonian. Most became extinct at the end of the Devonian but a few persist to the present day. Being jawless, most were either suction or filter feeders on mud or on planktonic organisms, although some may have used tooth-bearing sucker-like mouths to rasp away encrusting algae or feed on the bodies of other large organisms.

Subclass Heterostraci

The most primitive agnathans which had external armour but lacked paired fins. They lived in marine and fresh waters from the Cambrian to the Devonian (figure 52).

Subclass Osteostraci

These are known from Ordovician to Devonian rocks as forms which were either armoured or scaly but frequently possessed paired fins. The superficially eel-like members of the living Order Petromyzontida (the lampreys) derive from this subclass. Fossil lampreys are known from the Carboniferous of Mazon Creek (United States of America). The remaining classes of vertebrates are usually considered together as the Gnathostomata — the jawed vertebrates. Several of the anterior gill-supports of the agnathans have become modified as jaws which enable the gnathostomes to catch, hold and feed on large organisms, instead of feeding by filtering or rasping.

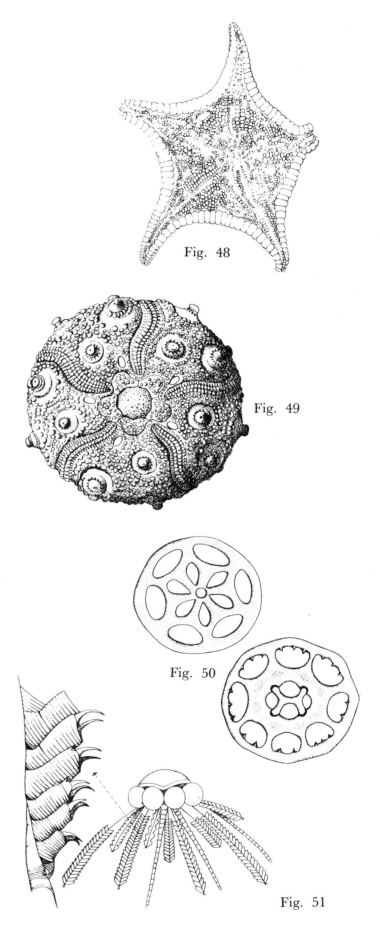

Fig. 48

Fig. 49

Fig. 50

Fig. 51

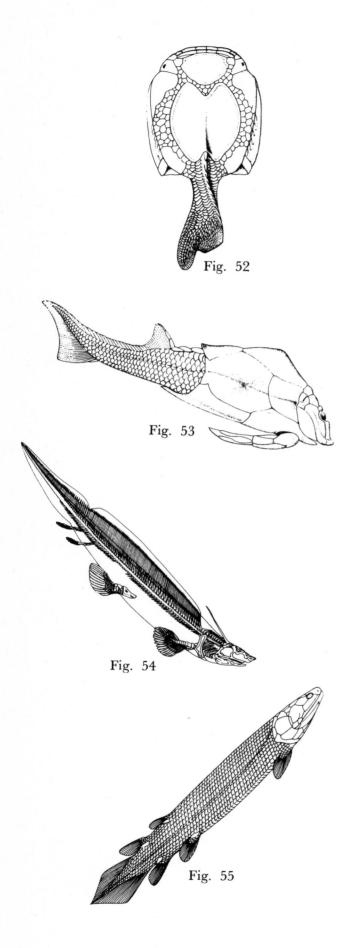

Fig. 52

Fig. 53

Fig. 54

Fig. 55

Class Placodermi

These extinct marine and freshwater fish-like vertebrates are known only from Devonian rocks. They were the earliest gnathostomes and had external armour, jaws and anterior and posterior paired fins (figure 53). The jaws and fins enabled some to become agile predators on other vertebrates. The largest placoderm was *Dunkleosteus* in which the head and fore-parts of the body were covered by dermal plates of bone, this armoured region being about 3 m long.

Class Chondrichthyes

The Chondrichthyes are characterized by a cartilage skeleton, fins with keratinous (horny) rays and tiny pointed scales resembling minute recurved teeth. They survive in abundance as the sharks, skates and rays. They first appeared in the Devonian and are mostly represented in the fossil record by their teeth, whole specimens being rare. Significant groups are the extinct Palaeozoic Cladoselachii and Xenacanthida (figure 54), and the living Selachii (sharks) which appeared in the late Palaeozoic, the Batoidei (skates and rays) which appeared in the Jurassic and the Holocephali (rabbit-fish) which also first appeared in the late Palaeozoic.

Class Osteichthyes

These are the bony fishes which are characterized by a bony internal skeleton, bony fin-rays and flat bony or keratinous scales. They are divided into the subclasses Actinopterygii which includes most of the 20,000 living fish species, and Sarcopterygii with only a few living representatives. Both first appear in Lower Devonian rocks and both were initially large diverse groups. Whereas the actinopterygians remain abundant to the present day, very few sarcopterygian lineages survived the Palaeozoic. Some palaeontologists also include in the Osteichthyes, a group called the Acanthodii or "spiny sharks", Palaeozoic forms characterized by possession of spines instead of fins. The Sarcopterygii includes the coelacanths, one of which, *Latimeria chalumnae*, survives in deep water off the Comoro Islands near Madagascar; and the Dipnoi or lungfish, a few species of which survive in freshwater habitats in South America, Africa and Australia. Both groups were more diverse in the late Palaeozoic than subsequently. The Sarcopterygii also includes the extinct rhipidistians (figure 55), closest relatives of the terrestrial vertebrates.

Class Amphibia

The amphibians were the earliest land vertebrates, being known from the Upper Devonian of Greenland as the fossils *Ichthyostega* and *Acanthostega*, the members of the Order Ichthyostegalia. They are clearly four-legged land vertebrates but *Ichthyostega* retains fish-like fin-rays in the tail and a fish-like articulation between the head and the backbone. In the Carboniferous and Permian, a wide range of primitive amphibians existed placed in the orders Temnospondyli (figure 56), Anthracosauria, Microsauria, Nectridea and Aïstopoda, but they declined during the late Permian and Triassic and only a few lineages of temnospondyl and microsaur descent survived to give rise to the living orders Anura (frogs and toads) and Urodela (newts and salamanders) in the early Mesozoic. Although the Anthracosauria had no amphibian descendants, they are believed to have given rise to the reptiles early in the Carboniferous.

Class Reptilia

Reptiles are more specialized than amphibians for a terrestrial existence. Unlike most amphibians, they have lost the aquatic larval phase of their life-history and, unlike all amphibians, they lay shelled amniotic eggs. The amniotic egg is characterized by the presence of a complex arrangement of internal membranes, in particular the amnion, which permits gas exchange between the embryo and the air outside the egg. These eggs are laid on land and, together with the reptiles' waterproof keratinous scaly skin, permit reptiles to live independently of standing or running water.

The earliest fossil reptiles are known from Upper Carboniferous rocks and are believed to be most closely related to the anthracosaurian amphibians. They diversified through the late Palaeozoic and Mesozoic and eventually filled a wide range of niches for terrestrial, aerial and aquatic vertebrates. A massive extinction of reptiles took place at the end of the Mesozoic and, although there are still over five thousand species of reptile alive today, they are no longer very diverse structurally, most living forms being small lizards and snakes. The Reptilia may be divided into five subclasses, three of which still survive. These subclasses are characterized on the basis of the construction of the skull, in particular the number and position of the so-called temporal fenestrae. These are openings in the bony cheek region of the reptile skull. Their edges act as insertion surfaces for the biting and chewing muscles inside the skull and their number and position relates to the reptile's feeding mechanism.

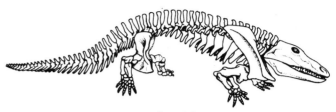

Fig. 56

Subclass Anapsida

These are the most primitive reptiles which retain the fish and amphibian condition of having no temporal fenestrae. They include the earliest known reptiles, the small insectivorous captorhinomorphs of the Carboniferous and Permian. They also include the Chelonia — the tortoises and turtles — represented by over two hundred living species and many hundreds of fossil forms from the Triassic to the Pleistocene. The heavy bony carapace of turtles is a very durable structure and is frequently preserved in freshwater and marine deposits (figure 57).

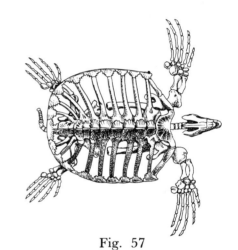

Fig. 57

Subclass Lepidosauria

These reptiles are characterized by two temporal fenestrae on each side of head, one above the other. They have various types of teeth but none have socketed teeth like the archosaurs described below. They first appear in the Permian as the Order Eosuchia, which gave rise to several groups which survive to the present day, namely the Rhynchocephalia (diverse in the Mesozoic but today represented only by the lizard-like Tuatara in New Zealand); the Lacertilia — figure 58 — (the true lizards first appearing in the Jurassic and abundant up to the present); and the Ophidia (the snakes, first appearing in the Cretaceous and diversifying through the Tertiary). One major extinct lizard family is the Mosasauridae, a family of large marine lizards up to 10 m long found in the Upper Cretaceous throughout the world.

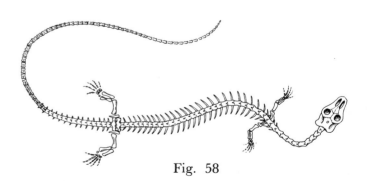

Fig. 58

Subclass Archosauria

The archosaurs have two temporal fenestrae on each side of the skull, like the lepidosaurs, but they are also characterized by the presence of teeth set in sockets and by a further opening on each side of the skull

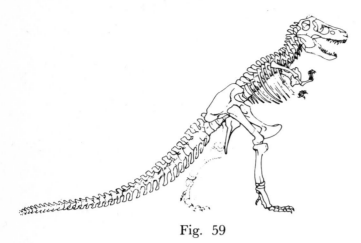

Fig. 59

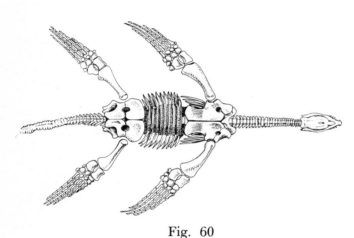

Fig. 60

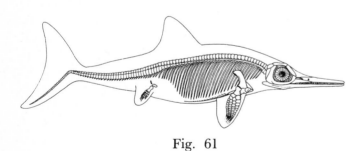

Fig. 61

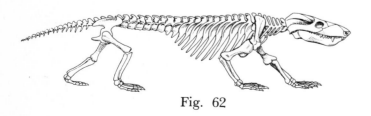

Fig. 62

between the nostril and the eye-socket. They appear in the late Permian and the basal archosaurian order, the Thecodontia, gave rise in the Triassic to several well-known orders including the Crocodilia which still survive; the Pterosauria; the birds, usually placed in a separate class Aves; and the two groups of dinosaurs, the Saurischia and the Ornithischia. The two groups of dinosaurs are characterized not ony by the large size of most of their members but by their "upright" limb posture (figure 59), the legs being held vertically under the body instead of being splayed out as in lizards and crocodiles. This is believed to be associated with more efficient forms of locomotion and may have been one of the reasons for the success of the dinosaurs.

Subclass Euryapsida

A completely extinct group of reptiles, characterized by one temporal opening on the upper cheek on each side of the skull. They first appear as small lizard-like terrestrial forms in the Permian but diversified as large marine forms through the Mesozoic, becoming extinct at the end of the Cretaceous. They include the Triassic nothosaurs and placodonts, the Jurassic and Cretaceous plesiosaurs and pliosaurs which grew to 10 m long (figure 60), and the ichthyosaurs which lived throughout the Mesozoic. The ichthyosaurs were the most specialized aquatic forms, convergent with sharks and dolphins in their general shape and undoubtedly completely aquatic. They had reevolved both dorsal fins and tail fins and avoided the necessity of laying eggs on land by giving birth to live young, as we know from the *Stenopterygius* specimens preserved after dying whilst giving birth to offspring (figure 61).

Subclass Synapsida

These forms had a single temporal opening set low down on the cheek on each side of the skull. They are popularly called "mammal-like reptiles" and represent the stem-lineage from early reptile to mammal. They first appear in the Upper Carboniferous and last appear in the Jurassic, overlapping slightly with the earliest mammals. Through the Permian and Triassic, the synapsids show progressive acquisition of the skeletal characteristics of mammals. The most advanced synapsids, the cynodonts (figure 62), had mammal-like teeth (differentiated into incisors, canines and cheek-teeth with several cusps) and mammal-like limbs but were still small-brained. The earliest mammals appear to have been dwarf cynodonts, perhaps small nocturnal shrew-like forms. Associated with their reduction in size is a rearrangement of some of the skull bones, namely the quadrate and the articular. These form the jaw hinge in synapsids (the quadrate on the skull and the articular on the lower jaw) whilst in mammals they are interposed between the stapes and the ear-drum next to the jaw-hinge and serve to transmit sound from the ear-drum to the inner ear.

Class Aves

Birds are rarely preserved as fossils since their skeleton is delicate and, as a result, their early evolution is poorly known. The oldest known bird is *Archaeopteryx lithographica* (figure 63), the Upper Jurassic form which possessed feathers but retained teeth and a long bony tail. Its skeleton resembles those of contemporary small cursorial dinosaurs and many palaeontologists believe *Archaeopteryx* and the later birds to be related to the dinosaurs.

The remaining birds are placed in the Subclass Neornithes. There are several early sea-birds, still toothed, known from Cretaceous rocks (e.g. *Enaliornis* from the Lower Cretaceous of Cambridgeshire), but only in the Tertiary do we see a wide range of birds in the fossil record. However, much of their diversification may have taken place in the Cretaceous.

Class Mammalia

Mammals (figure 64) are the most complex vertebrates. Living mammals may be differentiated from reptiles by such important anatomical and physiological characteristics as hair, sebaceous glands in the skin, homoiothermy (warm-bloodedness) and the ability to suckle the young by means of the mammary glands. However, these characteristics cannot be directly ascertained from fossil remains and a widely used but arbitrary skeletal character used to define fossil mammals is the position of the quadrate and articular bones mentioned earlier. Using this criterion, the earliest true mammals were small shrew-like insectivores of the Order Docodonta from the Triassic-Jurassic boundary. During the rest of the Mesozoic, they diversified into a range of insectivores, carnivores and herbivores, all small, including the triconodonts, symmetrodonts, multituberculates and pantotheres. These are all known from a very few good skeletons or skulls and many hundreds of isolated teeth. During the Cretaceous, the pantotheres gave rise to the marsupials and placentals, the two major groups of living mammals. After the extinction of the dinosaurs, these two groups diversified and, particularly the placentals, ultimately filled most of the niches for large terrestrial and aquatic vertebrates. A few major groups of placentals have become extinct during the Tertiary, particularly in South America, but most have at least some living representatives.

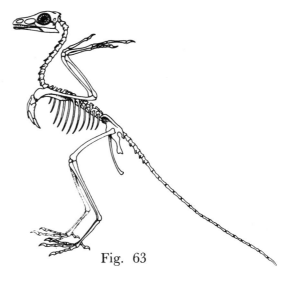

Fig. 63

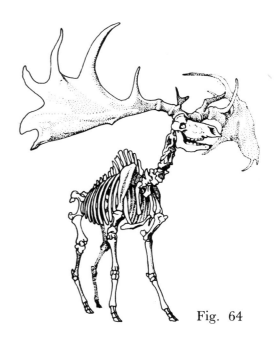

Fig. 64

Glossary

Aeolian used to describe a deposit accumulated by the action of wind, such as a desert sand-dune.

Allochthonous used to describe fossils which were not buried in or near the environment in which they lived but which had moved or been transported from elsewhere.

Amniotic egg egg characteristic of reptiles, birds and primitive mammals, with a hard or leathery shell and with several inner membranes forming fluid-filled sacs which protect and nourish the embryo.

Anthracosauria order of Carboniferous and Permian amphibians including *Seymouria* and the 4 m-long *Eogyrinus*, the largest Carboniferous amphibian.

Articular bone forming the hinge region of the lower jaw of fishes, amphibians, reptiles and birds. In mammals, it has become the malleus in the middle ear.

Aurochs the extinct wild ox, *Bos primigenius* found in Eurasia during the Pleistocene.

Autochthonous used to describe fossils which have been buried in or near to the environment in which they lived.

Belemnite extinct type of cephalopod resembling the modern squid but with a bullet-shaped internal shell, commonly found in Jurassic and Cretaceous rocks.

Benthic used to describe organisms living on the sea-bed, either in shallow or deep water.

Biocoenosis an assemblage of organisms which live together as a natural ecological community, or a fossil assemblage derived from such a community. Also called a life assemblage.

Breccia rock composed of coarse angular fragments of older rocks, mixed with finer material.

Carapace the bony shell of a tortoise or turtle, in particular the upper half.

Chalcedony form of quartz or silica with extremely small crystals. Agates are composed of this material.

Chitin a complex organic substance which forms the major component of the horny outer layer or cuticle of insects, spiders, scorpions and trilobites. In crustaceans it is hardened by lime-salts.

Clastic used to describe sediments or rocks accumulated from particles of material from elsewhere.

Cotyledon a simple leaf forming part of the embryo in seeds of plants. May be used as food-storage organ in the seeds of some plants. Monocotyledonous plant seeds have one, dicotyledon seeds have two.

Cursorial used to describe animals which are specialized for fast running.

Desmostylia an extinct group of aquatic herbivorous mammals of the Miocene and Pliocene, rather like manatees (seacows) but with prominent tusks.

Diagenesis physical and chemical changes occurring in a sediment (after deposition) which convert it to a solid rock, or similar changes occurring around a skeleton or shell after burial which turn it into a fossil.

Furcula anatomical term for the "wishbone" of birds. A central bone at the front of the shoulder girdle of birds, made up of fused clavicles.

Fusulinid an extinct planktonic protozoan of the family Fusulinidae, with a calcareous (chalky) shell shaped like a grain of wheat. Important zone-fossils in the late Palaeozoic.

Glyptodont one of a group of extinct large mammals from the Tertiary of South America. Glyptodonts were covered in bony armour and looked like gigantic mammalian tortoises up to 3 m long.

Heterochronism evolutionary change in structure brought about by different organs or parts of the body developing at different rates.

Labyrinthodont used to describe a group of amphibians of the Carboniferous, Permian and Triassic, characterized by large conical teeth with a labyrinthine structure; i.e., with complex infoldings of the dentine and enamel forming strengthening struts within the tooth.

Lacustrine used to describe anything associated specifically with lakes.

Lamellibranch a member of a large class of aquatic molluscs with a two-valved shell (hence also called bivalves); e.g., oyster, mussel, cockle.

Lamprey eel-shaped jawless vertebrate of the order Agnatha. One of the most primitive living vertebrates found in marine and fresh water in north and south temperate regions.

Marsupial one of a subclass of living and fossil mammals occurring now in Australia, South America and North America. The young are born at an early stage of development and then carried externally, often in a pouch; e.g., opossum, kangaroo, koala.

Mixosaur the most primitive type of ichthyosaur, found in the Triassic of Spitzbergen.

Moraine an accumulation of mud and stones deposited at the base of a glacier. Moraines left over from the last ice age are widespread in Northern Europe.

Multituberculate one of an extinct group of late Mesozoic and early Tertiary mammals which were apparently the first small herbivorous true mammals. They were replaced by the rodents in the Palaeocene.

Nektonic used to describe animals which live in open water, either in the sea or in large lakes, and which swim actively, thus being able to move independently of the currents.

Nummulites a genus of marine foraminiferan with a lens- or coin-shaped limy shell up to 3 cm across, found particularly in Eocene rocks. *Nummulites* was one of the largest of the unicellular animals, most being microscopic.

Oldowan Pebble Culture the earliest recognizable stone-tool culture found in Olduvai Gorge, Tanzania, along with early hominid fossils. The stone tools are only slightly shaped.

Onychophora small living class of tropical rain-forest invetebrates of the genus *Peripatus* which resemble soft-bodied millipedes and may be primitive relatives of millipedes and insects. A marine form *Aysheia* is known from the Cambrian.

Ornithischia an order of dinosaurs in which the pelvis has a three-pronged construction like that of birds. Includes a wide range of herbivores such as *Iguanodon*, *Stegosaurus* and *Triceratops*.

Pantothere one of an order of Jurassic and Cretaceous insectivorous mammals which are believed to have given rise to the marsupials and placentals.

Peduncle the stalk of a sedentary animal such as a crinoid.

Placental one of the major subclass of living and fossil mammals in which the embryo develops to an advanced stage within the uterus, attached to the maternal tissues by a complex structure called the placenta. Most living mammals belong here.

Planktonic used to describe passively floating organisms, mostly microscopic plants (phytoplankton) or animals (zooplankton) living in the open sea, generally, though not always, near the surface.

Pneumatized bone type of bone which is hollow and with one or more openings. Found in birds and pterosaurs. In birds, slender sac-like extensions of the lungs enter these bones and fill the hollows.

Pyrite a mineral composed of iron sulphide and sometimes called iron pyrites or fool's gold. It has a brassy metallic appearance and may be crystalline. Forms on or in fossils produced in acid, iron-rich sediments.

Pyroclastic general term used to describe material ejected by volcanoes such as tuffs, pumice, ash and gases.

Quadrate bone in the skull forming the hinge surface against which the lower jaw articulates in all jawed vertebrates except mammals where it functions as an earbone and is called the incus.

Rhipidistian type of fish belonging to the Sarcopterygii, such as *Eusthenopteron*. The rhipidistians occurred in the late Palaeozoic and are completely extinct. They are believed to include the forms which gave rise to the amphibians.

Sacral vertebra(e) bone(s) of the backbone which attach to the pelvic (hip) girdle.

Saurischia order of dinosaurs including the carnivorous dinosaurs, such as *Allosaurus*, and the gigantic sauropods, such as *Diplodocus*. Characterized by a four-pronged pelvis which is crocodile-like in construction.

Sauropod one of the group of gigantic long-necked dinosaurs of the Jurassic and Cretaceous such as *Diplodocus* or *Brachiosaurus*. They were probably canopy browsers like giraffes.

Schist a type of crystalline metamorphic rock produced by high temperature and pressure within the earth's crust.

Sclerite one plate or unit of an external skeleton of an invertebrate such as an arthropod or an echinoderm.

Symbiosis an association of two or more dissimilar organisms to their mutual advantage.

Symmetrodont one of a group of Jurassic or Cretaceous mammals, apparently small insectivores.

Tectonic used to describe anything associated with movements of the earth's crust from earthquakes to continental drift.

Temnospondyl one of an order of Carboniferous to Triassic amphibians which varied from small salamander-like forms to large crocodile-like forms. They probably gave rise to the living frogs and toads.

Thallus simple vegetative plant body showing no structural division into roots, stems or leaves. Found in living seaweeds, lichens, liverworts and many early land-plants.

Thanatocoenosis a group of organisms brought together only after death, or the resultant accumulation of fossils. Also called a death assemblage.

Theca external skeleton surrounding the body of certain types of marine invertebrate, usually sac-like, e.g., in crinoids.

Thecodont used to describe an order of Upper Permian and Triassic diapsid reptiles with teeth set in sockets. Gave rise to the crocodiles, dinosaurs, pterosaurs and birds.

Therapsid member of the most advanced order of mammal-like reptiles (synapsids). The therapsids are of Upper Permian to Middle Jurassic age and include the cynodonts and *Lystrosaurus*.

Triconodont one of a group of small Mesozoic mammals, including the earliest true mammals.

Ungulate a hoofed mammal. Nearly all are relatively large herbivorous mammals including the living pigs, deer, sheep, antelope, horses, rhinoceros and elephant, and the fossil mammoths and *Baluchitherium*.

Index

(page numbers in italic refer to illustrations)

A
Acanthodes 96
Achistrum 124
Acitheca polymorpha 120
Aeger 153, 156–7
Aepyornis 13, 188
Agnatha 118, 209
algae, blue-green *see* Schizophyta
algae, brown *see* Phaeophyta
algae, green *see* Chlorophyta
algae, red *see* Rhodophyta
Allosaurus fragilis 148–9
Alps 166, 169, 170
Amaltheus 140–1
amber *36*, *37*, 42
ammonites *32–5*, 46, *48–9*, 68, 76, 80, *88–9*, 120, 134, *140–1*, 148, 150, 207
amphibians 87–92, *114–5*, 122, 132, 210
Anancus arvernensis 182
Anaplotherium commune 14
Anapsida 211
Anatosaurus 42
Andes 166, 169
Annelida 203
Angara 112
angiosperms 142, 144, 148, 170, 200
Antarctica *74*, *75*, *100*, 108
Antennata 205
Anthozoa 202–3
Antrimpos noricus 60
Apennines 169–170
Arachnida 205
Archaeocypoda veronensis 174
Archaeocyatha 201, *202*
Archaeopalinurus levis 60
Archaeopteryx lithographica 41, 82, 87, 92, *93*, 148, *212*
Archaeozoic era 48, 87
Archimedes 94
Archosauria 211–12
Arctica islandica *180–1*, 186
Arduino, Giovanni 97
Aristotle 14
Arnioceras 48–9
Arthropoda 204
Askeptosaurus italicus 133
Aspidorhynchus acutirostris 161
aurochs 190
Australia *74*, *75*, *100*, 108, 148
australopithecines 173, 192
Avicenna 15
Aysheaia pedunculata 68, 204

B
Bacillariophyta 197
Baluchiterium 24, 172
bacteria 196
barnacles 71
belemnites *86*, 135, 148, 150, 207
bennettitales 144, 199
Besano *133*, 134
birds 130, 148, 150, 172, 212
Blastoidea 207–8
Bos primigenius 190
Bothriolepis canadensis 82–3
Brachiopoda *32–4*, 56, 67, 107, *126*, 203
Branchiosaurus 123
Bryozoa *94*, 203
Burgess Shale 42, 118

C
Calymene blumenbachii 108
carbonification 36, 39–40
Cardium 50
Caucasus 169
Cephalopoda 134, 148, 172, 206
Cene 60, 136–9
Cenozoic era 48, 100, 164–177
Ceresiosaurus calcagnii 81
Cerithium 51
cetaceans 172
Chaetognatha 203
Chelicerata 205
Chirotherium prints 71, *85*
Chlorophyceae 197
Chlorophyta 197
Chonchodon 13
Chondrichthyes 210
Chondrites 66
Chordata 209–213
Chrysophyceae 196
Chrysophyta 196
Ciliata 201
Coccolithophoridae 196
Coelenterata 202–3
Coelophysis bauri 20
Coleia 153
Colonna, Fabio 16
condensation (palaeontological) 52, 66
conifers 134, 144, 200
Conodonts 204
Conularia 202
corals *47*, 134, 170
Cordaitales 199, *200*

D
Dactylioceras athleticum 88–9
Dactyloidites 105
Dalmanites caudatus 108
Darwin, Charles 18
Dasycladiaceae 198
dating of rocks 46, 48–50
deer, giant *188*
Desmostylus 173
diatoms *197*
Diatryma 172
Dibranchia 207
Dicotyledoneae 200
Dimetrodon 127
dinoflagellates *197*
Dinophyceae 197
Dinornis maximus 183
dinosaurs 20, 130, 142, 144, 146, *147*, 150–1, 172, 212
Dipleurozoa 202
distillation 40
Dollo, Louis 84
Dolomites 134
Dolothoceras sociale 122
Drepanaspis gemuendensis 113
Drepanosaurus unguicaudatus 137

Crinoidea *95*, *130–1*, *135*, 208
Crioceras 140
Crustacea *24*, *25*, 135, *153*, *166–7*, 205
Crystoblastus melo 106
Cuvier, George 10, 16, 18
cycadales 134, 144
Cycadophytae 199
Cycas 134
Cymatium 170
Cymatophlebia longialata 158–9
Cyphosoma koenigi 54–5
Cyrtospirifer *126*
Cystoidea 207

E
Echinodermata *35*, 84, 134, *158*, *169*
Echinoidea 208
Edaphosaurus 127
Edrioasteroidea 208
Eichstätt 148, 155, 158, 162
elephant, evolution of 172
Empedocles of Agrigentum 14
Eohippus 172
Eosuchi 133

equiseta 134
Eryon arctiformis 24–5
Eudimorphodon ranzii 138–9
Euryapsida 212
Eurypterida 205
Eusthenopteron 87, 88

F
ferns *32–33*, 134, 198–9
fish *17*, 88, *138*, *152*, *160*, *161*, 209
Flabellipecten flabelliformis 58
Flood, The 15–16
Foraminifer 201
fossils, facies 68
fossils, trace 70–1
fossils, zone 46, 48
Fusinus 170

G
Gadoufaouà 146–7
gastropods 11, *32–4*, *41*, *51*, 206
Giza 11
ginkgoales 144, 200
Ginkoo biloba 200
glaciations 68–70, 101, 180, 182–5
Glossopteris 75
Gondwanaland 75, 112, 199
Gotland 118
graptolites *40–1*, *111*, 117, 208–9
Greenland 150
Gymnospermae 199–200
Gyroptychius 114–5

H
Haeckel 82
Halysites catenularia 107
Herodotus 10, 11
heterochronism 84
Himalayas 100, 166, 169
Holothurioidea 208
Holzmaden 22, 24, 28, 42, 144
Homer 12
hominids 173, 192
Homotelus bromidensis 87
horse, evolution of 80–82, 172
Hydrozoa 202
Hypselocrinus 95

I
ichthyosaurs *22*, 135, 148, 150, 212
Ichthyostega 87, 88, 120
India *74*, *75*, *100*, 108, 148
insects *158–9*, 205
Italy 169–70

L
Labidosaurus hamatus 78–9
Labyrinthodontia 88, 127, 140
Laevaptychus 140
lamellibranchs 67, *130–1*, *180–1*, 206

Latimeria 88, 210
Lepidodendrales 198
Lepidodendron 118–9
Lepidosauria 140, 211
limestone *47*
Limulus 156, 205
link fossils 86–7
Linnaeus 195
lizards 140, 150
lumachella *48–9*
lycopods 134
Lycopsida 198
Lyme Regis 147–8
Lystrosaurus 75, 142

M
mammals 130, 166, 170, 172, 213
mammoth *18*, 26, 188
Mastodonsaurus 140
Mazon Creek *124–5*
medusae *105*, 202
Megaloceros 188
Mene rhombea 175
Merostomata 205
Merycoidodon 68–9
Mesodon macrocephalus 160
Mesolimulus walchii 156
Mesosaurus tumidus 75, 77
Mesozoic era 48, 99, 128–163
mineralization 36
mixosaurs 135, 148
moa *183*, 188
Moeritherium 17
molluscs *72–3*, 135, *170–1*, 186, 205–6
Monocotyledoneae 200
Monoplacophora 206
Monte Bolca 28, 170, 174–7
mosasaurs 150
mummification 42
Murex 170
Myriapoda 205

N
Narona 170
Nautiloidea 206
neoteny 84
Nereis 124
Neuropteris 124–5
nothosaurs *81*, *134–5*, 140, 144
nummulites 11, 201

O
ontogenesis 82
Onychophora 204
Oreopithecus 173
Ornithischia 144, 212
Orthis 107
Osteichthyes 210
Osteno 152–3
Owen, Richard 17

P
Pachypleurosaurus edwardsi 26–7
Palaeophonus nuncius 118
palaeontological condensation 52, 66
Palaeozoic era 48, 87, 98, 102–128
panama, isthmus 76
Pangaea *74*, 112, 116, 130, 144
Paraspirifer 126
Pentacrinus subangularis 22–3
Phacops rana 108–9
Phaeophyceae 197
Phaeophyta 197
Phorusrhacos 172
Pinna 67
Placodermi *82–3*, 114, 120, 210
placodonts 135, *136*, 144
plesiosaurs 140, 148, 150, 212
Pleurodictyum problematicum 58
Pliny the Elder 15
pliosaurs 148
Pliosaurus ferox 145
Po, River *188*, 189
Polychaeta 203–4
Polyplacophora 206
Porifera 201
Primary era *see* Palaeozoic
primates 173
Productus 116
Protoclytiopsis antiqua 60
Protomedusae 202
Protozoa 201
Psilophytales 198
Psylopsida 198
Pterichthyodes 114
Pteridophyta 198
pteridosperms *75*, 134, 199
Pterobranchia 208
Pterodactylus 28–9, *162–3*
Pteropsida 198
pterosaurs 92, *162–3*
Pycnogonida 205
Pyrenees 169
Pyrrophyta 197

Q
Quaternary era 178–192

R
Radiolaria 201
radiometric methods of dating 48–50
Ramapithecus 173
Rana pueyoi 65
Raphidonema farringdonense 142
reptiles 88–96, 123, 127, 130, 211–2
reptiles, flying 29, 130, *138–9*, 142, 144
reptiles, marine 27, 135, 140, 148
Rhodophyceae 197
Rhodophyta 197
Rhynchonella
Rhyncocephalia *16*, 211
Rocky Mountains 166, 169

S

Saccocoma pectinata 158
Sahara, rock carvings *192*
Sanmiguelia lewisi 200
Sarcopterygii 210
Saurischia 144, 212
Scaphopoda 206
Scheuchzer 16
Schizophyceae 196
Schizophyta 196
Schizomycetes 196
Scyphozoa 202
sea urchins *54*, 84, *168*, 208
Secondary era *see* Mesozoic
sedimentation 32–6, 46, 52
sedimentation, chemical 46
sedimentation, clastic 46
Senftenbergia plumosa 120–1
Seymouria 87, 92, 127
Seymouria baylorensis 84–5
Sigillaria 120, 198
Silicoflagellata 196
Smerdis minutus 88–91
Solnhofen *24*, 28, *41–2*, 92, 148, *154–63*
South Africa 75
South America 75, 76, 108
Sparnodus vulgaris 176–7

speciation 80
Spermatophyta 199–200
Sphenopsida 198
Spirifer 56
sponges *142*
starfish *169*
Stelleroidea 208
Stenarthron zitteli 158
Steno, Nicolaus (Niels Stensen) 16
Stenopterygius 22, 24
Stomochordata 208
Strabo 11
stratification, crossed 64
stratigraphic sequences 46
stromatolites *196*
symbiosis 55, 70
Synapsida 94, 127
systematics 194–213

T

Taeniaster 38
Tarsophlebia eximia 39
Teneré desert 12, *146–7*
Tertiary era *see* Cenozoic
Tethys *74*, 99, 112, 132
Theophrastus of Ephesus 15

Ticinosuchus 71
Tintinnidae 201
travertine *40*
Triadobatrachus 65
trilobites 56, 62–3, 70, 87, 108, 109, 118, 204
Trilobitomorphida 204
turtles 142, *143*
Tyrannosaurus rex 151

U

Uintatherium 172
Urals 98, 108
uranium 50

V

vertebrates 118, 209–213
Voltzia 134

W

Wadi Rum *110–1*
Walchia 134, *200*
Wapta, Mount 117
Wegener, Alfred 77
worms *66*, 203–4

This English-language edition first published 1984
© Burke Publishing Company Limited 1984
Translated and adapted from *Il Grande Libro della Preistoria*
© Vallardi Indstrie Grafiche S.p.A. 1981

All rights reserved. No part of this publication may be reproduced, stored in a retrieval system, or transmitted in any form or by any means, electronic, mechanical, photocopying, recording or otherwise, without the prior permission of Burke Publishing Company Limited.

Acknowledgements
The Publishers are grateful to Dr. Lucia Wildt, in association with First Edition, for preparing the English translation of the text of this book and to Dr. Andrew Milner for scientific advice.

The Publishers are also grateful to the following for permission to reproduce illustrations: Giovanni Pinna, Luciano Spezia, Arch. Vallardi, Bertarelli, Erwin Christian, Marka, Pedone, Pubbliaerfoto, The Tate Gallery and Titus. The drawings are by Gabriele Pozzi.

CIP data
Pinna, G.
  Prehistory.
  1. Man, prehistoric
  I. Title II. Il grande libro della preistoria.
  English
  930   CC100
  ISBN 0 222 00895 4

Burke Publishing Company Limited
Pegasus House, 116-120 Golden Lane, London EC1Y OTL, England.
Burke Publishing (Canada) Limited
Toronto, Ontario, Canada.
Burke Publishing Company Inc.
540 Barnum Avenue, Bridgeport, Connecticut 06608, U.S.A.
Typeset in Linotron 202 by
Graphicraft Typesetters Limited, Hong Kong
Printed in Italy by Vallardi Industrie Grafiche S.p.A.